D1685992

Farming, Fertilizers and the Nitrate Problem

by

T.M. Addiscott, A.P. Whitmore and D.S. Powlson

Rothamsted Experimental Station
Harpenden
UK

C·A·B International

C·A·B International   Tel: Wallingford (0491) 32111
Wallingford   Telex: 847964 (COMAGG G)
Oxon OX10 8DE   Telecom Gold/Dialcom: 84: CAU001
UK   Fax: (0491) 33508

First printed 1991
Reprinted 1992

A catalogue record for this book is available from the British Library

ISBN 0 85198 658 7

Typeset by Leaper & Gard Ltd, Bristol
Printed and bound in the UK by Redwood Press Ltd, Melksham

Contents

Preface   v

1. Nitrate: Perception of the Problem   1

2. Nitrogen in Crop Production   15

3. The Physics and Chemistry of Nitrate Leakage   31

4. The Biology of Nitrate Leakage   44

5. Measuring Nitrate Losses from Soil   55

6. Chasing Nitrate with a Computer: Modelling   73

7. Sources of Nitrate Leakage: Arable Farming   92

8. Sources of Nitrate Leakage: Grassland and Organic Farming   110

9. Some Complications   128

10. Ways of Decreasing Nitrate Leaching   138

Bibliography   157

Index   165

Preface

This book has been written in response to increasing fears that nitrate from agriculture has found its way into drinking water causing cancer, cyanosis in infants, the growth of toxic algae in rivers and seas, and untold mayhem in otherwise balanced, natural ecosystems. What is worse (so we are told) is that much nitrate has not yet had these effects but is waiting to do so, moving unseen in underground waters, sinking slowly, insidiously and inexorably towards our taps. In short, a nitrate time-bomb.

Nitrate fertilizer is blamed for these fears, both perceived and apparent, because quantities of nitrate fertilizer applied to crops world-wide have been increasing at the same time as nitrate levels in drinking water or water in underground aquifers. A statistician would say that the two are well-correlated. Statistics, however, might be described as the branch of science that tells us exactly how inexact our measurements are. It is thus always open to a greal deal of misunderstanding and unfortunately, manipulation. Any statistician worthy of his profession would immediately point out that because he says nitrate concentrations in drinking water are well-correlated with fertilizer applications it does not mean that the fertilizer is to blame. It would after all be ridiculous to assert that the rise in nitrate concentrations has caused an increase in fertilizer applications. Yet this is all that correlation can tell us: that the two have occurred simultaneously.

There is an old and well-worn correlation between the numbers of bottles of whisky sold around the turn of the century and the number of clergymen in England and Wales at the same time. No doubt a few alcoholic churchmen contributed to the sales of whisky, but the most sensible interpretation of the data is that as distilleries expanded to serve the needs of a growing population so did the church.

And farming? The issue is complicated else why should we write this book, but our central theme is that the rise in nitrate concentrations in aquifers and drinking water has increased both in line with the population and as farming has intensified to feed the increased numbers of people. We shall show that there are other aspects to intensive farming besides fertilizer use that have led to nitrate leaching.

In Chapters 1 and 2 we discuss the fears that the public and press have of nitrate, what it might do to us and what it does not do. We look too, at the way in which plants use nitrate, why they need it and what proportion of the nitrate they receive they turn into edible or saleable products. In Chapters 3 to 5 we look at how nitrate behaves in soil, where it comes from and where it goes. A key question addressed in Chapter 5 is how we know these things: how we measure nitrate. Building on this, we show in Chapter 6 how scientists have made mathematical models of the key processes in the nitrogen cycle and how these have led to an increased understanding of the nitrate problem despite its complexities. In Chapters 7 and 8 come the main content of the book as we look at how much nitrate leaks from the different systems of agriculture, arable and annual crops grown for seed or leaf or root, how much leaks from animal systems and what part organic farmers play in all this. Then at the end of the book we turn to other sources of nitrate in the environment, explore issues other than drinking water quality before setting out the measures that we believe will quickly and most effectively bring about a reduction in nitrate in drinking water.

We believe that most of this information is needed for a complete understanding of the nitrate problem, but we also understand that among our readers are those who come armed with prior knowledge and those who need to address only a single aspect of the whole problem. Engineers seeking an understanding of the broad issues may not really want to know how and why plants use nitrogen. And ordinary members of the public may not wish to understand the processes that go to make up a mathematical model of nitrate leaching. We certainly encourage these people to be eclectic, and have written the book so that each chapter may be read alone – by someone with prior knowledge of the basic principles. On the other hand we have had lay people in mind when writing and can assure them, that if they take each chapter in turn, we shall lead them carefully through the complexities of the nitrogen cycle to reach a greater understanding of just why they find nitrate in their drinking water.

Chapter One
Nitrate: Perception of the Problem

Adequate food and clean water are properly regarded as basic human rights. People in industrialized societies, however, have grown used to more than basics. Food has become plentiful and cheap, while our taps yield unlimited supplies of water suitable for drinking or cooking. Both have become expectations where 45 years ago they were no more than hopes for the future in countries devastated by war and stricken by famine. These expectations are beginning to conflict (House of Lords Select Committee, p. 7, 1989). The root of this conflict lies in the particular properties of a very commonplace naturally-occurring chemical entity called nitrate. Like many substances, nitrate is essential to life, but a nuisance and possibly a hazard in the wrong place at the wrong time. The same is true of water. Nitrate dissolves in water with great readiness and is found in most natural waters – in rain, in rivers and lakes, in the sea and importantly in water stored in porous rocks such as chalk and sandstone. Our tap water may come from any of these sources, in some cases even the sea. One reason for the interest in nitrate, and thence for this book is that concentrations of nitrate in these natural waters have been increasing steadily for the past 30 years. This increase has coincided with increasing use of nitrogen fertilizer by farmers as shown in Figure 1.1(a–c). Another very important reason for the interest in nitrate is that the European Community (EC) has recently imposed a limit of 50 milligrams per litre (mg l^{-1}) for nitrate in potable water (Council of the European Communities, 1980). We need to draw attention here to a possible source of confusion. The EC limit refers to 50 mg l^{-1} of nitrate, but when we wish to relate *nitrate* in water to losses of *nitrogen* from the soil (in Chapters 7–10 for example), it is more helpful to think of the nitrogen in the nitrate without the associated oxygen, that is, N rather than NO_3. The divisor for converting a nitrate concentration to a concentration of nitrogen from nitrate is 4.427, and the EC's 50 mg l^{-1} limit for nitrate becomes an 11.3 mg l^{-1} limit for nitrate-nitrogen.

It is clearly not prudent to allow too much of any chemical substance to accumulate in our water supplies without asking questions about its

Chapter One

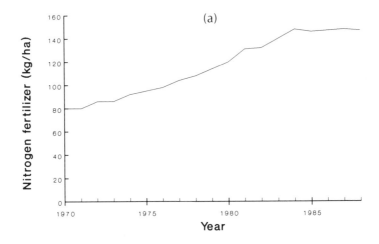

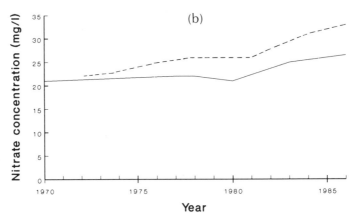

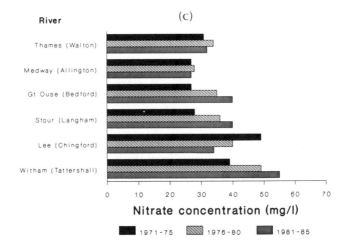

consequences. This is particularly so in the case of water supplies stored in porous rocks (aquifers) that are used and replenished over long time scales. We need to ask how the nitrate gets there, what it does, and how we can best stop it from getting there. These are all questions that are amenable to rational discussion and sensible solutions. Because the increase in nitrate concentrations in water has *coincided with* the increase in the use of nitrogen fertilizers it is assumed to have been *caused* by it. It follows from this that making farmers cut their use of nitrogen fertilizer should alleviate the problem. And, as the argument goes, this will also help to demolish the massive mountains of grain that subsidized farmers in Europe are building at the public's expense. In short, we can kill two birds with one stone. But can we? Is it as simple as that?

Farmers are the most convenient people to blame for the nitrate problem, and the grain mountains, but other people are involved, such as politicians, industrialists and even voters. History has played its part too. The Second World War left Europe in a dreadful state and it was designated a famine area in 1947. Older British readers will probably remember ration cards and National Dried Egg and similar delicacies. The slogan 'Dig for Victory' used during the war in the UK could well have become 'Dig for Survival'. Food production had to be increased, so old grassland was ploughed as it had been in the war (Fig. 1.2) and the British Government introduced a subsidy on the use of fertilizers that was to last until 1966. As we moved from the 1940s into the 1950s it was no longer food that was in short supply. Expanding industries, the car industry for example, needed labour and this came from the agricultural work-force (Fig. 1.3). Fewer men on the land meant that those who remained had to be more productive. Machines replaced the missing men, herbicides replaced hoes and chemical fertilizers from bags increasingly replaced the traditional manures because they could be handled more easily and applied by machine.

As farming became more efficient it became more profitable and land became attractive as an investment for large institutions such as pension funds. Land prices rose (Fig. 1.4) and with them the need for consistently good yields of crops to repay investment. Larger applications of nitrogen fertilizers were part of a package of measures needed, a trend that was exacerbated by the sharp rise in interest rates during the 1970s. Perhaps the biggest influence of all, however, has been the Common Agricultural

Fig. 1.1. (a) Increase in the amount of nitrogen fertilizer (kg ha^{-1}) applied to all crops and grass 1970–88 (Survey of Fertilizer Practice); (b) increase in the concentration of nitrate in water drawn from two bore-holes between 1970 and 1986; Mill Meece (–), Hatton (. . .); (c) change in the concentration of nitrate in water drawn from some UK rivers between 1971 and 1985 (House of Lords Select Committee on European Communities, 1989).

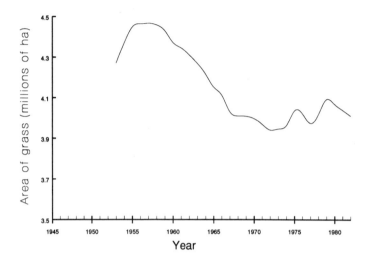

Fig. 1.2. Decline in the area of permanent grassland in UK 1950–82 (Marks, 1989).

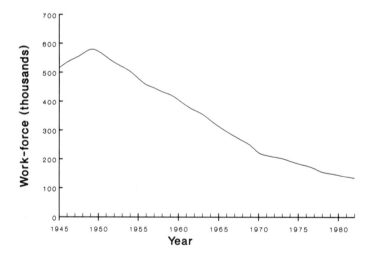

Fig. 1.3. Decline in the agricultural work-force of the UK 1945–82 (MAFF, 1947–90).

Policy (CAP) of the European Communities (EC). Because of the way in which land is inherited in some European countries many people have an indirect interest in farming, or farm on a part-time basis. Small farmers vote in large numbers and politicians in some European countries cannot afford to ignore their interests. It is worth noting in passing that Hitler's rise to power depended to no small extent, on support from many disgruntled

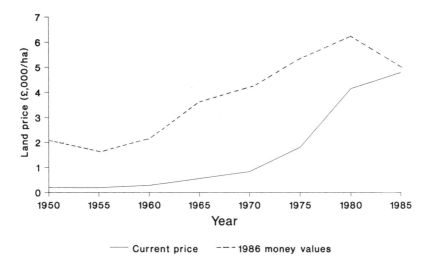

Fig. 1.4. Increase in the price of land (Marks, 1989).

farmers. Maybe this memory prompted the EC to introduce the CAP, which basically ensured that farmers were assured of a market and a consistent price for their produce even when there was a glut. The intent was to protect farmers against the wild fluctuations in the prices they received for their produce, but one result has been the 'mountains' and 'lakes' that have been so roundly damned in the press. The excess produce must, in many cases have been the result of the use of extra nitrogen fertilizer.

Politicians have been very ready to castigate 'greedy farmers' for both grain mountains and nitrate pollution. It seems clear, however, that the problems have no simple origin and that both arise from the problems, activities and aspirations of a great many people, including politicians. Blaming either problem simply on nitrogen fertilizer could be a massive over-simplification, so is it logical to force farmers to cut their use of nitrogen fertilizer? The answer is important for several reasons. The first is that cuts in the use of nitrogen fertilizer mean cuts in production and thence declining earnings from agriculture and probably fewer jobs in rural areas. This may be offset if the farmer is compensated for loss of income, as under the Nitrate Sensitive Areas scheme in the UK for example. Another reason why the question is important is that nitrogen fertilizer increases crop yields with relatively little cost to the farmer; take it away and food prices must rise in the shops. A third reason is that decreasing food production at home means more imports from abroad which affects the balance of payments and diverts food that may be needed elsewhere. There has been a famine at least somewhere in the world almost every year since the war.

It would be disastrous if cuts in the use of nitrogen fertilizer and other restrictions on farming caused the problems listed above without making any impact on the nitrate problem. It would be even worse, however, if our failure to do anything about the nitrate problem led to a further increase in nitrate concentrations at a later date and caused a large-scale catastrophe. We have to get to grips with the problem. When we look at the science of the nitrate problem we find that trying to over-simplify it is no more sensible than trying to over-simplify the 'political' origins of the nitrate problem. If however, you tackle the problem one part at a time, as we have tried to do in this book, even the more complex aspects can be seen to have their own logic and therefore to offer solutions.

Nitrate – a hazard, a nuisance or both?

We shall be discussing the chemistry of nitrate in Chapter 3 but first we need to introduce one important point now. This is that nitrate has a close relative called nitrite. Writing down the formulae for potassium nitrate, KNO_3 and potassium nitrite, KNO_2, shows that the difference between them is that the nitrite has one less oxygen atom than nitrate. This results in appreciable differences in the chemical and biological behaviour of the two substances. What we find in the present context is that nitrate is a nuisance when in the wrong place in the environment but not in itself a threat to our health, whereas nitrite is definitely a potential health hazard. Once again, however, this depends on it being in the wrong place at the wrong time; in other circumstances it can save life, for example as a preservative in salted meat where it prevents the growth of *Botulinas*, the organism that causes botulism. Another key point to remember is that nitrate can be changed into nitrite by microbes requiring oxygen. This can happen in the soil, in water and in our bodies.

Nitrate problems in the environment

Nitrogen is essential for plants to grow – we will expand on this in Chapter 2. Nitrogen fertilizer helps the farmer to grow better crops, but it also means that nitrate out of place encourages plants to grow where they are not wanted, notably in rivers and lakes (The Royal Society, 1983; UK DoE, 1986). Reeds may grow to excess, thus narrowing waterways and possibly overloading and damaging banks, while under the surface of the water other plants foul the propellers of boats and also infuriate anglers. Water supply conduits become clogged and machinery damaged. Large plants are not the only problem. The algae are very small single-celled plants that grow on the surface of practically anything, including water. They are not noticeable until they grow to excess, when they cause algal 'blooms', arguably better described as algal scum, that we now see to a worrying extent on rivers and lakes. These blooms are not only unsightly. When they die, the bacteria that consume them use oxygen to do so; this lessens the supply of oxygen to other living organisms such as fish, which may die as a result. The whole ecological balance of the river or lake may change and we can lose species that we would like to see there. Such effects of nitrogen enrichment are termed 'eutrophication'.

Nitrate is not the only culprit. Algal blooms depend not only on nitrate but also on phosphate, which is often the factor initiating their growth. This phosphate could come from agricultural land, but it is usually held by the particles of soil and is not lost nearly as readily as nitrate. Phosphate also gets into natural water from sewage, where it may originate from human excreta or from the phosphatic detergents that we use so freely in our homes. We are to blame as well as the farmer.

Nitrate and health

Nitrate itself is not toxic. If you found it mentioned in a medical textbook it could well be because sodium nitrate and ammonium nitrate are used to treat phosphatic kidney stones in humans. Doses of up to 9 g day^{-1} have been used for this purpose without any adverse effect – other than to the kidney stones. Nitrate becomes a problem only when it is converted into nitrite. When that happens we have to consider two problems, the 'blue-baby' syndrome and stomach cancer.

The 'blue-baby' syndrome or methaemoglobinaemia, can occur when children less than about 1 year old consume too much nitrate. Microbes in the stomach convert nitrate to nitrite and when this reaches the blood-stream it reacts with the haemoglobin, the agent that transports oxygen around the body. Normal oxyhaemoglobin, which contains iron in the ferrous form, becomes methaemoglobin in which the iron is in the ferric form. As a result, the oxygen-carrying capacity of the blood is lessened. Very young children are susceptible because foetal haemoglobin, which has a greater affinity for nitrite than normal haemoglobin, persists for a while in the bloodstream, and because their stomachs are not acid enough to inhibit the microbes that convert nitrate to nitrite. This is too complex a process to discuss further here other than to say that the unfortunate infant suffers what might be described as 'chemical suffocation', a very severe condition that can prove fatal. We can perhaps make this point most effectively by quoting from the doctors who reported the last death from methaemoglobinaemia in the UK. 'There were diarrhoea and vomiting and the child's complexion was slate-blue' (Ewing and Mayon-White, 1951). In a similar but fortunately non-fatal case, 'Blood drawn from a vein was a deep chocolate-brown' (Ewing and Mayon-White, 1951).

Happily the 'blue-baby' syndrome, or cyanosis, is extremely rare. The last death it caused in the UK was in 1950 (Ewing and Mayon-White, 1951) and the last confirmed case here in 1972. A world-wide survey (Sattelmacher, 1962) showed that in all only about 1060 cases had been published although many others had clearly gone unreported. Others have occurred since, notably an outbreak in Hungary where 1353 cases occurred between 1976 and 1982 (Deak, 1985). Early authors reporting this condition used the term 'well-water methaemoglobinaemia' (Comly, 1945) and the death mentioned above was associated with water from a well (Table 1.1). The wells involved in cases in this and other countries have often been privately dug wells, which may have been badly positioned in relation to the disposal of human or animal excreta. This is suggested by the fact that the well-water involved in many cases was contaminated by bacteria as well as containing too much nitrate.

Table 1.1. Some cases of methaemoglobinaemia and associated nitrate concentrations in water.

Source	Country	Cases	Nitrate concentration (mg l^{-1})
Comly (1945)	USA	2 anecdotal	388, 619 283–620
Choquette (1980)	USA	1	1200
Busch and Meyer (1982)	USA	1	545
Johnson *et al.* (1987)	USA	1	665
Ewing and Mayon-White (1951)	UK	2	200*, 95(P)
Acheson (1985)	UK	14	>100
Deak (1985)	Hungary	95 1258	40–100 >100
Hye-Knudsen (1985)	Denmark	1	200(P)

All of these cases were associated with water from wells. The wells were dug privately in 98% of the cases.
*Known fatal case.
(P) Water known to have been bacterially polluted.

It is not easy to decide what constitutes an excessive concentration of nitrate in water, because this depends on whether or not the water also contains any bacteria and on the overall health of the young consumer. The water responsible for the death in 1950 contained about 200 mg NO_3 l^{-1} and traces of nitrite. That in a non-fatal case which happened in the same year had 95 mg l^{-1}, but was also 'heavily contaminated with coliform organisms'. Picking a 'safe-level' from the variety of evidence available is not easy, but three august bodies, The Medical Research Council, The Institute of Biology and The Institute of Cancer Research recently stated to a committee of the House of Lords that the majority of cases have occurred when the water contained more than 100 mg NO_3 l^{-1}.

In Britain we like to blame everything on the weather, and the 'blue-baby' syndrome need be no exception. The fatal case in 1950 and the non-fatal case mentioned above, which happened in the same year, were both associated by the doctors reporting them with 'heavy rainfall following 12 months of exceptional drought'. We hope that the significance of these factors will be clearer to you when you have read the rest of this book. You should note, however, that they are significant only because the water involved came from wells. We do not know any cases of the 'blue-baby' syndrome associated with tap water from the mains supply.

Stomach cancer has also been associated with the concentration of nitrate in potable water. There are good theoretical reasons for proposing this association. Nitrite produced from nitrate could react in the stomach with an organic compound called a secondary amine coming from the breakdown of meat. The result of their union (Fig. 1.5) would be an N–nitroso compound. Unfortunately, N–nitroso compounds can cause cancer because they are capable of modifying certain of the components of DNA. This mechanism is only a hypothesis and we need to test it further against the facts. The result is not as clear-cut as would be ideal but it is fairly definite.

In one study a group of doctors from the Radcliffe Infirmary at Oxford identified two areas of the UK in which the incidence of stomach cancer was unusually high and two in which it was unusually low (Forman *et at.*, 1985). These showed very clear differences; the mortality from stomach cancer in the high-risk area was roughly double that in the low-risk area (Table 1.2). People attending the hospital as visitors who were not ill themselves were asked to give samples of saliva. According to the hypothesis the

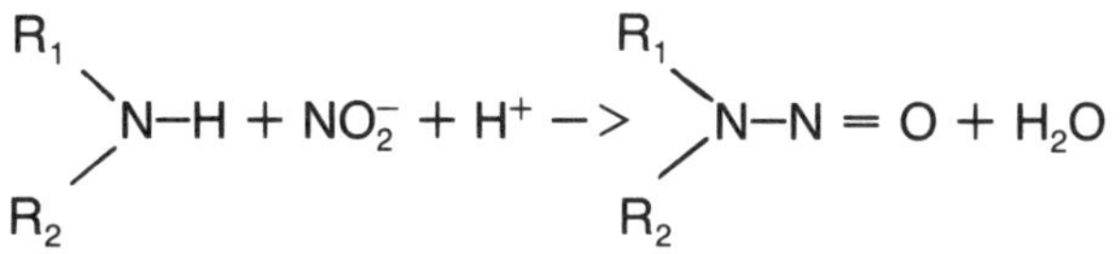

$$\begin{array}{c} R_1 \\ \diagdown \\ \diagup \\ R_2 \end{array} N{-}H + NO_2^- + H^+ \longrightarrow \begin{array}{c} R_1 \\ \diagdown \\ \diagup \\ R_2 \end{array} N{-}N = O + H_2O$$

Fig. 1.5. Formation of a N–nitroso compound.

Table 1.2. Nitrate concentrations in saliva samples from people living in areas of high and low risk with respect to stomach cancer (from Forman *et al.*, 1985).

	Low risk		High risk		Both low	Both high
	Area I	Area II	Area III	Area IV	I + II	III + IV
All samples	208	157	107	108	–	–
Refined samples*	172	150	97	117	162†	106†

*Refined sample: the donor had not eaten or drunk in the 2 h before sampling.
†There is less than one in 10 000 probability that this difference happened by chance.

samples from the high-risk areas should have contained more nitrate and nitrite than those from the low-risk areas. The reverse proved to be so; the samples from the low-risk population had nitrate concentrations 50% higher than those from the high-risk population. Another study (Beresford, 1985) looked at the nitrate concentrations in drinking water in 229 urban areas in the UK between 1969 and 1973, and at deaths from stomach cancer in the same areas at the same time. According to the hypothesis there should have been a positive relationship between them. In fact there was negative one.

These results look fairly conclusive but we need to ask ourselves if they could be misleading for any reason. We need to think in particular about two possible factors, other forms of nitrate consumption and the possible lag between consuming nitrate and the development of stomach cancer.

Leaf vegetables contain considerable nitrate. A lettuce we bought recently had a concentration of 300 mg nitrate l^{-1} in its sap, 7.5 times more than the local water supply that afternoon. A substantial proportion of the nitrate we consume, often at least half, comes from food rather than water. The exact proportion depends on where we live and what we eat. This means that the first (Oxford) study quoted could have been dominated by the nitrate the participants had eaten rather than the nitrate they had consumed in tap water. If this was true we would not need to worry so much about nitrate in water. A further complication is that leaf vegetables contain vitamin C, which is reputed to be an anti-cancer agent. It certainly is an anti-oxidant, which means that it hinders the oxidizing action of nitrate. When nitrate oxidizes something nitrite is formed. If the people in the various areas had differed appreciably in their consumption of leaf vegetables this could have confounded the results. When the authors of these two studies took account of social and other possible confounding factors they found that the relationships between nitrate and stomach cancer were less clearly negative but still showed no evidence at all that nitrate concentrations in water determined the incidence of stomach cancer.

Some cancers grow with alarming rapidity but many grow only slowly and may become evident only many years after they were triggered. This means that relating stomach cancer *now* to nitrate in water *now* may not be relevant. There may be a time lag. A study of 43 local authority areas in Kent attempted to take account of a possible lag by relating deaths from stomach cancer in the period from 1959 to 1973 to the concentrations of nitrate in water in 1946 (Clough, 1983). Here there was a significant positive relationship between stomach cancer and nitrate for males but not for females. Taking account of the time lag thus seems to alter the picture but as before we need to check whether these results could be misleading for any reason. There are three possible problems. One is that we do not know how reliable the data for nitrate concentrations in water were, remembering that this was in 1946 just after the end of the war. Another is that the areas studied were small relative to the areas in the more recent studies; this means that the death rates would have been affected more by variations that occurred by chance. The third problem is that the author was not able to take account of the social factors that were found to make the results of the other studies less certain. These results should make us cautious in drawing conclusions from the other studies, but do not necessarily invalidate them.

If nitrate does cause stomach cancer, one group of people should be strongly affected – those who work in fertilizer factories and are exposed to dust containing nitrate. Workers in such factories do have larger concentrations of nitrate in their saliva than workers who are in comparable jobs but not exposed to nitrate dust. They do not, however, suffer a great incidence of stomach cancer. This has been shown in two separate studies, one of which (Al-Dabbagh *et al.*, 1986) is summarized in Table 1.3.

We have found very limited evidence to support the hypothesis linking cancer of the stomach with nitrate in water. We started saying that 'Nitrite produced from nitrate could react in the stomach. . . .' What we did not say was how the nitrite was produced from nitrate and this was simply because there is no obvious way in which it should be produced. The adult stomach is too acid for the conversion to occur there. This leaves the mouth as the only possible reactor. Chewing vegetables thoroughly might possibly give enough time for microbes in the mouth to effect the change on the nitrate they contained. But what about drinking a glass of water or a cup of tea? Could the microbes really intercept the nitrate as it passed through the mouth? This conversion from nitrate to nitrite seems to be the weak link in the hypothesis. If this link is not really there we would expect to find that

Table 1.3. Mortality from cancer of the stomach and other causes among male workers in a nitrate fertilizer plant, 1 January 1946 to 28 February 1981. Number of deaths observed (obs) among those heavily and less heavily exposed to nitrate compared with the number of deaths that would have been *expected* (exp) from local population statistics (from Al-Dabbagh *et al.*, 1986).

Disease	Number of deaths					
	Heavily exposed		Less heavily exposed		Both groups together	
	obs	exp	obs	exp	obs	exp
Stomach cancer	7	7.22	5	4.84	12	12.06
All cancers	59	51.36	32	35.47	91	86.83
Respiratory diseases	21	30.97	15	20.07	36	51.04
Ischaemic heart disease	56	67.64	36	45.98	92*	113.72
All causes	193	219.78	111*	148.33	304**	368.11

Nitrate and nitrite concentrations in saliva increased with exposure to nitrate in a survey of current workers at the plant.
*There is less than one in 20 probability that this difference between observed and expected happened by chance. ** There is less than one in 1000 probability that this difference happened by chance.

consuming nitrate would be quite safe but that consuming nitrite would not. This is just what was shown in a large-scale Canadian study (Risch *et al.*, 1985). The risk of stomach cancer was negatively associated with nitrate intake, but if the anti-cancer effect of the vitamin C in the vegetables was taken into account the negative association became no association at all. On the other hand, the cancer risk was definitely associated with the intake of nitrite, coming mainly from preserved meat to which it is added. This emphasizes the point that it is nitrite that is the hazard. Should it therefore be left out of preserved meat then as increasingly it is? Not necessarily unless another preservative is added instead because nitrite is the most reliable way of preventing botulism.

The Canadian study showed clearly where the main risk lies but it does not solve the issue. It did not attempt to assess the influence of nitrate or nitrite ingested in water, and it does not take account of any lag between the triggering of stomach cancer and its appearance as a medical problem.

When we consider all the evidence we are left with a slight uncertainty about the time-lag effect but a strong general impression that there is no real link between stomach cancer and nitrate in water. This impression is heightened by the fact that while nitrate concentrations in water have been increasing during the past 30 years the incidence of stomach cancer has been declining. The impression seems to be shared by the British Government which issued a statement in the mid-1980s accepting that there was no link between stomach cancer and nitrate in water (Acheson, 1985).

This assembly of medical information leaves plenty of scope for debate as to how much nitrate in water is 'safe'. The European Commission has, however, made such a debate redundant in the countries over which it has influence by imposing a limit of 50 mg NO_3 l^{-1} in domestic water supplies (Council of the European Communities, 1980). The World Health Organization sets two limits, a 'recommended' limit of 50 mg l^{-1} and a maximum of 100 mg l^{-1} (22.6 mg l^{-1} as nitrate–nitrogen). The choice between these figures makes a big difference to the percentage of the land in certain areas that can be kept in arable cultivation (see Table 1.4 and its note). The decision by the EC to adopt 50 mg l^{-1} as the limit can be defended as giving a safety factor of 2 relative to the 100 g l^{-1} deemed safe by the three medical authorities quoted above, but it has unfortunate consequences for those who farm in the drier areas of England. It also adds to the burden on those working in the water supply industry. To meet the limit they have to blend water from different sources, which is difficult because the nitrate concentrations fluctuate, or else remove the nitrate from the water by chemical or microbiological means as we shall see in Chapter 9. The idea of

Table 1.4. Effect of maximum permissible nitrate concentration in the drainage water on the proportion of land that could be used for arable cropping*

Maximum permissible nitrate concentration in drainage water (mg l^{-1})	Proportion of land that could be used for arable cropping
50	27
60	33
70	40
80	46
90	52
100	58

*Assuming nitrate concentration in the draining water is the average (166 mg l^{-1}) for arable cropping in the UK and the remaining land is used for rough grazing (taken from: House of Lords Select Committee, p. 2439, 1989).

removing the nitrate by something like a giant water-softener that takes out nitrate instead of calcium (lime) sounds logical. The problem is that these exchange resins as they are properly called, have to be regenerated with brine, like ordinary water-softeners. Brine contaminated with nitrate then has to go somewhere where it will not cause further problems. Putting it in the sea sounds sensible – unless the water-works are far from the sea. But in any case the sea is no place for surplus nitrate; algal blooms on the sea are increasing and nitrate may bear more responsibility for them, relative to phosphate than it does for fresh-water blooms. Given the proper chemical additions to stored water, microbes will convert nitrate to gases; unfortunately the chemical additions to the water are not without their effect on human metabolism and one of the gases formed (nitrous oxide) is a serious atmospheric pollutant (Chapter 4).

Clearly, both stopping nitrate getting into water and getting it out again are very complicated matters with no easy or readily acceptable solutions. A further aspect of the problem must cause frustration to all involved. This we describe as the 'Cleopatra syndrome'. Cleopatra was seen by her critics as an extravagant hussy because she bathed in ass's milk. What, we wonder, would those critics have made of modern industrialized societies where we spend great effort and sums of money on keeping our water safe for making up baby's bottle – and then not only bathe in it but wash the car with it and flush the WC with it, not to mention sprinkling it on the garden so that further nitrate dissolves in it!

Chapter Two
Nitrogen in Crop Production

A short history of the study of plant nutrition

Man has understood the importance of soil since the beginnings of agriculture. It was realized early on that we could not go on taking nourishment from the soil indefinitely without allowing it to recuperate. What was not realized was exactly what it was that the crops were taking from the soil or what the process of recuperation involved; however it was known that allowing the natural forest to recolonize the land made it fruitful again and that growing crops in certain sequences or rotations was also helpful. Only relatively recently, in Victorian times, however, did the scientific basis of soil 'recuperation' or 'replenishment' become understood fully.

Among the early scientists who wanted to know the origin of plant matter were van Helmont and Boyle in the 17th century. Van Helmont made very careful experiments with soil in pots, in which he weighed the pot, the soil and the seed at the start and recorded all the water given. At the end he measured the gain in weight of the plant and loss of weight of the soil (Table 2.1). On the basis of the evidence he had he attributed the plant's gain in weight to water and decided that the change in weight of the soil was too small to be of significance. He missed two key contributors to the growth of the plant. One was the carbon that the plant obtained from carbon dioxide in the atmosphere by photosynthesis; this was not surprising, because carbon dioxide had yet to be discovered. The other was what we call the mineral nutrients – nitrogen (in the form of ammonium or nitrate), phosphate, potassium, calcium, magnesium, sulphate and others.

Van Helmont was partly right. Water is by far the largest component in the live tissues of plants and many other organisms, but something has to hold the water in place. All living matter is composed of cells and cells cannot be built without proteins, which are composed principally of carbon and nitrogen. The growth of plants and other organisms occurs as cells divide and regrow, a process that needs energy. Without this knowledge,

15

Table 2.1. Balance sheet of gain in weight by a willow tree planted by Van Helmont. Quoted by Russell (1973).

	Start	End	Gain/loss
Weight of soil	200 lb	199 lb 14 oz	0*
Weight of willow	5 lb	169 lb	164 lb

*It is unfortunate that Van Helmont did not realize the significance of this small loss. The gain of the 164 lb in the willow tree indeed came from water and carbon dioxide (although he failed to realize the importance even of carbon dioxide) but the 2 oz loss from soil was real and not experimental error.

which scientists now take for granted, there was considerable confusion as to the origin of plant material that was not resolved until about 100 years ago.

In 1840 Justus von Liebig launched a scathing attack on the then accepted idea that plants obtained carbon from soil; he asserted, quite rightly, that plants obtained carbon from the air and other nutrients from the soil. Liebig was wrong in some of the details of his ideas, but he played a key role in laying the foundation of the more rigorous approach to plant nutrition we have today. Liebig's advance left one major controversy to rumble on, that of where plants obtain the nitrogen they need to grow. Liebig knew that plants got their carbon from carbon dioxide in the air, despite the fact that carbon dioxide makes up less than 1% of the air. He concluded, not unreasonably, that since nitrogen makes up nearly 80% of the air plants must draw their nitrogen from the air. This, however, is not so, and Lawes and Gilbert, the founding fathers of the Rothamsted Experimental Station, were able to prove that many plants could *not* get nitrogen from the air. One of the reasons why the controversy lasted until the mid 19th century was that certain plants, such as clover and peas and beans (the leguminous plants), consistently added nitrogen to the soil, nitrogen that could only have come from the air. Was it possible that some plants could obtain nitrogen from the air while others could not? And why was it that the plants that could apparently get nitrogen from the air failed to grow without nitrogen from soil when the soil was sterile? It was only when the new science of microbiology evolved some years later that the questions could be answered. It was not the plants themselves that drew the nitrogen from the air but a special group of microbes in the soil, now known as *Rhizobia*. These microbes live in association with the plant's roots in nodules formed around the root. They are able to 'fix' nitrogen from the air and supply it to the plant in return for carbohydrate and other foods and the protection provided by the root nodule. We shall return to this in Chapter 4.

Nitrogen – the key to crop growth?

Nitrogen is vital to the growth of plants. It is part of all the essential constituents of cells: the chlorophyll that is essential for photosynthesis, the nucleic acids, DNA and RNA in which the pattern for the plant's growth and development are encoded, the proteins which include the enzymes which catalyse all biochemical processes, and, not least, the walls that hold the cell together. Nitrogen-containing compounds are involved in practically all the biochemistry of life.

Nitrogen is not the only vital nutrient. Phosphorus has a major role in the building of nucleic acids and in the storage and use of energy in cells. Equally, potassium plays a key part in the transport of carbohydrates from where they are produced in the leaves to where they are used in other parts. It also helps plants to use water efficiently and to maintain their osmotic and electrical balance. Magnesium is central to the function of chlorophyll, while sulphur is a constituent of some key proteins. Calcium is important in structural tissue and in other ways, and a number of other elements have key roles albeit in very small amounts – they are known collectively as the 'trace elements'. Why then is nitrogen of such particular importance to the farmer?

All the nutrients increase the growth and yield of crops but nitrogen has by far the largest effect, except for crops which form large storage organs such as potato tubers; these are strongly influenced by the supply of phosphorus and potassium as well. For wheat, barley and other cereal crops nitrogen is the main key to yield. Crops that have insufficient nitrogen are stunted, yellowish and sickly-looking. Even a small amount of nitrogen given at the right time relieves these symptoms, and larger amounts can increase the yield of grain by a factor of five or six on some occasions. Within the range in which nitrogen increases yield the farmer will get roughly 20 kg of extra grain for each kilogram of fertilizer nitrogen that he supplies. In the UK nitrogen costs 30–40 p kg^{-1} and wheat grain fetches 10 p kg^{-1}, so the farmer gets a handsome return on every 30 p invested. He also gets a better return on his investment in other resources for the crop: seed, pesticides, machinery and most notably, the land, on which he may be paying a high rate of interest. The latter resources constitute 'fixed' costs – the farmer pays them regardless of the yield he gets – so he has every incentive to use a cheap nutrient to boost his yield and thence his profit. It is not only the farmer who benefits; nitrogen fertilizer helps to keep down the price of food in the shops.

The grain to fertilizer nitrogen ratio of 20 kg kg^{-1} described in the last paragraph does not go on indefinitely, for reasons that will be described later, and there is a point, described as the 'optimum' at which applying extra nitrogen fertilizer ceases to be worthwhile. The extra nitrogen is largely wasted and it is likely to contribute to nitrate pollution. The farmer has good reasons for not wanting to apply less than the optimum amount of nitrogen, so he tends to over-supply it as a form of insurance. This insurance policy can have a premium that is costed in terms of nitrate pollution, so it is important that his advisors can give him an estimate of the amount of nitrogen fertilizer needed by the crop in which he can have so much confidence that no insurance is necessary. We shall return to this point later.

Nitrogen – the key to yield

Nitrogen fertilizer influences the yields of crops in four distinct ways. Two involve the quantity of yield and the third its quality, while the fourth could be said to concern the quality of life of the crop. The principles tend to be the same for all crops, so we shall outline the mechanisms involved and then consider their net effect for crops grown in various parts of the world.

1. Leaf area. Nitrogen may make leaves larger, and therefore able to trap more light, by increasing both the rate at which cells multiply – and thence their number – and also the size of each cell. It also increases the number of leaves, and plants vary in their strategies for achieving this. Cereal plants, such as wheat and barley, and the grasses to which they are related have the habit of producing extra shoots, known as 'tillers'. These probably evolved as a means of exploiting favourable growing conditions and many tillers die if nutrients are in short supply. Adding nitrogen increases the number of tillers that survive; extra tillers mean extra leaves, extra stems and more yield. Plants such as potatoes with a more complex leaf structure tend to produce extra branches and therefore extra leaves in response to nitrogen. Nitrogen also helps to extend the duration of the plant's leaf cover. It is very mobile in plants and as the older leaves die they release their nitrogen to benefit growing tissue; thus adding extra nitrogen delays the overall dying back of the leaves and maintains a greener, leafier plant. We have to take care, however, not to encourage vegetative growth at the expense of the growth of the parts of the plant that we wish to harvest; giving too much nitrogen to potato plants, for example, delays the initiation of tubers.

2. Crop development. We saw above that injudicious use of nitrogen can delay the development of the potato tubers, but some research suggests that with cereal crops skilful timing of applications of nitrogen fertilizer can encourage the development of flowers and thence the yield of grain. It can also seemingly cause more tillers to be fertile and set seed, but in practical terms it is difficult to distinguish this effect from the effect of nitrogen on tiller production.

3. Crop quality. Concentrating too much on yield can be a mistake, because yield and quality are not always compatible and the latter may be the more important in some circumstances. Bakers need flour that is milled from grain rich in certain proteins and nitrogen fertilizer can often, but not always, be used to increase the protein in grain. Pasta makers also need high-protein grain but the proteins that they require are different from those wanted by bakers. Brewers need barley to make malt but they need grain with as little protein as possible so nitrogen needs to be used very cautiously. The yield of sugar beet is increased by nitrogen but the concentration of sugar in the beet decreases as more nitrogen is given while that of certain unwanted amino–nitrogen compounds goes up. The sugar industry therefore pays premiums for high sugar concentrations and low amino–nitrogen concentrations, and the farmer has to balance yield and premiums when working out how much nitrogen to supply. Rape is a crop that is grown for the oil in its seeds. Nitrogen increases the yield of seed but may decrease the concentration of oil in it, so a balance has again to be made. With potatoes the increase in yield has to be set against the fact that nitrogen can produce too many over-large potatoes, or potatoes with too large a water content, a matter of concern to the manufacturers of potato crisps.

4. Side effects. Fertilizer nitrogen is not an unmixed blessing to crops. The extra growth of leaves that it brings means extra weight for the stems to bear and extra resistance to wind and therefore stress on the stems. This can increase transpiration and so the use of water by the plant. The nitrogen enlarges the cells of stems as well as leaves, making them more succulent but mechanically less strong. One consequence is that cereal crops become more likely to suffer 'lodging', that is, being laid flat by the wind. This sorry sight will probably be familiar to older readers. It has become less frequent with the introduction of modern short-strawed varieties and of chemicals that regulate stem growth but the farmer still has to keep lodging in mind. Another problem is that the more succulent stems are more prone to attack by insects and fungal diseases that have to be controlled. Rape crops tend to have broadly similar problems. Their increased succulence can also make it more attractive to pigeons; pigeons are a major factor in rape production and will feature later in the story.

Applying nitrogen fertilizer: how much?

It should be clear from the above that judging the correct amount of nitrogen fertilizer to supply a crop is like a balancing act: give too little and you can cut your yield, give too much and you risk lodging – and, as we shall see later, you increase the risk of nitrate pollution. How can the 'optimum' application be estimated?

The response curve

The 'classic' way of determining the optimum is to make a field experiment in which the crop of interest is grown on plots in a field that receives various (perhaps six or seven) different applications of nitrogen as fertilizer. The different applications are usually replicated to even out the variability in the field. If the crop is a cereal the yields of grain are plotted against the amount of nitrogen supplied (Fig. 2.1) to give the response curve. If the crop is sugar beet the yield of sugar is plotted and so on. Figure 2.1 shows three of the commonest forms of 'curve'. Curve (b) is the most common: as more nitrogen is supplied the increase in yield per kilogram becomes smaller; this is known as the diminishing returns curve. If the soil already contains a lot of nitrogen that is available to the crop the 'curve' may show a straight line decline, indicating that giving nitrogen cuts

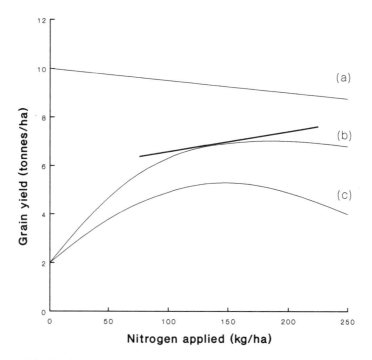

Fig. 2.1. Idealized curves of the response of the yield of grain (winter wheat) to applied nitrogen fertilizer: (a) linear decline; (b) diminishing returns; (c) catastrophic.

the yield [curve (a)]. There is also what could be described as the 'catastrophic' curve, in which the extra nitrogen causes a steady increase in yield up to a point at which there is a very sharp down-turn in yield [curve (c)]; this could for example represent lodging. A negative or a catastrophic response curve usually shows fairly unequivocally how much nitrogen can usefully be applied, but the smooth 'diminishing returns' curve is amenable to a variety of interpretations – all of them somewhat arbitrary. The most common criterion for making decisions nowadays tends to be that of financial return, and the optimum that is usually calculated is therefore the economic optimum. This is obtained [the tangent to curve 2.2(b)] by examining the change in the ratio of extra grain to extra nitrogen fertilizer and finding the point at which the extra grain no longer pays for the extra fertilizer. This very commonly made calculation is unsatisfactory for several reasons.

1. It depends on the price of grain and the price of fertilizer both of which vary considerably. The price of wheat increased greatly over a very short period in 1972, so the economic optimum increased sharply too. It is notable that up to the mid-1970s the quantity of nitrogen applied as fertilizer to British winter wheat crops was on average less than the quantity of nitrogen estimated to have been removed in crops (Fig. 2.2). This desirable state of affairs was reversed in 1975–76. The arbitrariness of the economic optimum also shows when different nitrogen fertilizers are considered.

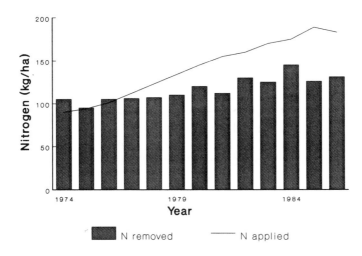

Fig. 2.2. Comparison between the amounts of fertilizer applied to winter wheat in the UK 1974–86 and the amount of nitrogen removed in grain during the same period of time (Sylvester-Bradley *et al.*, 1987`

Urea is cheaper per unit of nitrogen than ammonium nitrate (at the time of writing). The economic optimum is therefore greater for urea, by quite a margin, than for ammonium nitrate, which is nonsense from a physiological point of view.

2. Because crops tend to become more prone to attack by disease or insects as more nitrogen is given them, the farmer needs to spend more on chemicals to protect the crop. These extra costs ought to be included in the calculation of the optimum, but they are frequently omitted.

3. Is the economic optimum still relevant anywhere? In Western Europe and North America it is over-production of food rather than under-production that is the problem and the main concern is with the environment. This is a luxury that most countries do not enjoy, but even in countries where food production is at a premium we need to ask whether a blunt implement like the economic optimum is really what is needed.

One final problem must be mentioned before we leave the response curve and that is that the information from the curve is essentially retrospective. The optimum, economic or otherwise, is not known until about 6 months after the fertilizer has to be spread. It may be relevant to the same field the following year, but the weather may dictate otherwise. All that can really be done is to accumulate a set of optima and to try to relate them to factors that can be assessed before fertilizer has to be applied. Soil type and preceding weather are clearly relevant, and the residues left by the preceding crop are obviously important too.

The N-min method

A winter wheat crop that produces 8 t ha^{-1} of grain, a good but far from exceptional yield, probably removes about 200 kg ha^{-1} of nitrogen from the soil in grain and straw. Suppose the soil contains about 50 kg ha^{-1} of nitrogen available in mineral form (ammonium and nitrate) in April when the fertilizer has to be applied. A rather simple way of estimating the nitrogen application is to subtract 50 from 200 and say that the crop needs 150 kg ha^{-1} as fertilizer to 'top-up' what is in the soil. Several factors are omitted by this simple method of estimated crop requirement known as the 'N-min' method, notably the quantity of nitrogen already in the crop. However, the general principle of topping-up what is already there has been used quite widely in Western Europe (e.g. Dilz *et al.*, 1982). In general the optimum application of nitrogen fertilizer seems to have been quite well related to the N-min, the amount of mineral nitrogen in the soil. Mineral nitrogen means nitrogen present in the form of ammonium or nitrate. The result was that the optimum application topped up the N-min to give a total pool of about 250 kg ha^{-1}, except where the N-min exceeded this value on its own. One big advantage of the N-min system is that where it works successfully it goes a long way towards preventing unnecessary nitrogen fertilizer from being used. Unfortunately, one place where the N-min system has not been made to work satisfactorily is the UK except where large amounts of mineral nitrogen are expected in soil from, for example, applications of manure or slurry (Shepherd, 1990). The problem probably lies in the diversity of soils and weather conditions that we have, but a recent study at Rothamsted also suggests that it may be necessary to sample the soil more deeply than we have so far (Addiscott and Darby, 1991). It is the fact that the soil has to be sampled for analysis that constitutes the main disadvantage of the N-min method. Although sampling in stoneless sandy or silty soil is not a problem, sampling a heavy clay soil or a soil with numerous flints is another matter; there is much to be said for estimating N-min with a computer model.

The surplus nitrate curve

Feeding a crop is not unlike feeding a small child in that neither can tell you how much it wants. The child leaves unwanted food on the plate – or on the carpet; the crop leaves it in the soil. Several experiments testing various applications of nitrogen fertilizer to winter wheat have shown the pattern depicted in Figure 2.3 (e.g. Chaney, 1990). The amount of nitrate

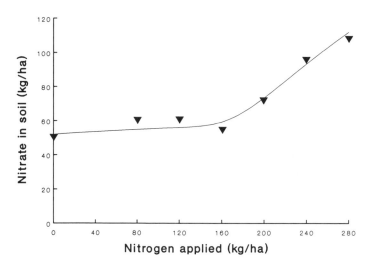

Fig. 2.3. Relationship between the nitrogen applied to winter wheat and the amount remaining unused in soil. The 'surplus nitrate' curve (Chaney, 1990).

left in the soil when the crop is harvested remains relatively small and roughly constant as more nitrogen is given as fertilizer up to a point at which the amount of nitrate left rises sharply. Thus this nitrate measurement gives a fairly clear indication of when the crop is satisfied, and therefore of the optimum. What is even more encouraging is that this 'point of satisfaction' often falls fairly close to the economic optimum when the latter is computed with a realistic grain:price ratio. The optimum computed from the surplus nitrate curve has two advantages over that computed from the yield-response curve. It remains uncontaminated by economics – no prices are involved – and it can probably be assessed better by statistics. Identifying the point at which the slope of the curve reaches a particular value on a smooth curve that is flattening out is not easy whether it is done by eye or statistics, and much may depend on the choice of equation used to compute the curve from the experimental points. It can happen that where two different equations are used to fit response curves to the same set of yields they suggest optimum nitrogen applications that differ by 40 kg ha^{-1}. If the surplus nitrate 'curve' consists of a horizontal line that turns upwards at the 'point of satisfaction', this point can be defined relatively easily as either the point at which a statistically significant change of slope occurs or the point at which the upward curvature becomes significant. The other great advantage of an optimum computed from the surplus nitrate curve is that surplus nitrate may pollute, so this optimum seems to have far greater environmental relevance than the economic optimum.

Applying nitrogen fertilizer: what kind?

The comparison made above between feeding a crop and feeding a small child can be extended; both have fairly definite preferences. For most purposes the crop has the choice between ammonium and nitrate as its source of nitrogen. Plants absorb nitrate readily and they are equipped with very efficient enzyme systems for converting nitrate to ammonium. In the first stage an enzyme called nitrate reductase converts nitrate to nitrite and the nitrite is then converted to ammonium by nitrite reductase.

$$NO_3^- \rightarrow NO_2^- \rightarrow NH_4^+ \qquad\qquad (2.1)$$

These twin enzymes are among the most important in the plant world. Further enzyme systems then convert the ammonium into amino acids and proteins and the process, generally described as nitrate assimilation, is intimately linked with the photosynthetic processes that assimilate carbon from the carbon dioxide in the air. Plants can also take up ammonium from the soil, thereby eliminating the need for converting nitrate to nitrite and nitrite to ammonium, but this is not always advantageous because of the link between nitrate assimilation and carbon assimilation. Ammonium can become toxic to plants in some circumstances, particularly when they are stressed by heat or shortage of water. In practice, as we shall see in the next chapter, ammonium added to the soil as fertilizer is usually converted rapidly to nitrate by nitrifying bacteria. The same is true of ammonium produced when other microbes decompose soil organic matter, and the ubiquity of nitrate reductase in the plant world presumably reflects the similar ubiquity of such nitrifying organisms.

The commonest nitrogen fertilizer in the UK is ammonium nitrate, which is sometimes supplied as prills (small spheres) and sometimes impregnated into chalk, giving a product called 'nitrochalk'. Ammonium nitrate is a material that needs to be treated with respect; mixed with aluminium powder it became 'Ammonal', one of the more powerful explosives used in the Second World War. It is treated with a stabilizer for use as a fertilizer, but it is not used much in areas where terrorist bombs are a problem. Ammonium nitrate contains 34.5% nitrogen, so nearly two-thirds of its weight has to be transported unnecessarily. Urea, which is 46% nitrogen, is more popular in large countries where transport is a problem. Urea is converted to ammonium in the soil by an enzyme called urease. Calcium nitrate and calcium ammonium nitrate are also used as fertilizer materials, and where a compound fertilizer containing nitrogen, phosphorus and potassium is needed, mono- or di-ammonium phosphate may be used to supply both nitrogen and phosphorus. The use of saltpetre (potassium nitrate) of course supplies both potassium and nitrogen.

Applying nitrogen fertilizer: when?

When we were discussing how much nitrogen fertilizer is needed we were thinking partly about optimizing the yield. The question of when to apply it is related to that of why we apply it. It is done to help the crop to exploit its full potential, and particularly to ensure that there is enough nitrogen available for this at all times. We also have to include the environmental objective of ensuring that there is as little surplus nitrate as possible in the soil at risk of leaching. Looking at the relationships in time between the amount of mineral nitrogen supplied naturally by the soil and the amount of nitrogen that should be in a winter wheat crop (Fig. 2.4) shows that in autumn the soil supplies enough to meet the needs of the crop and often more than enough. This would be true in most soils. During the winter the soil is cool and the microbes produce little mineral nitrogen, but the crop does not need much so there is no problem. During the spring the crop's growth accelerates and its demand for nitrogen eventually outstrips the

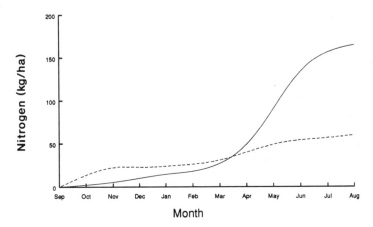

Fig. 2.4. Idealized diagram showing the demand of nitrogen from a winter cereal throughout its growth and the ability of the soil to supply nitrogen. Nitrogen mineralized (----); crop uptake (—).

soil's capacity to supply it. Exactly when this happens depends on the particular soil and crop, but this is the point at which the crop needs some nitrogen fertilizer, which may or may not be its main application. This point may not be reached until the crop makes its peak demand, which is at the time that the stem starts its main growth. A substantial application of nitrogen fertilizer is usually given at about this time.

Can all the nitrogen be given at the same time, or should it be divided into two or more smaller applications? Much depends on the soil's capacity to supply nitrogen in the early spring. A crop growing in a fertile soil may well manage quite comfortably until the stems shoot up and it needs its main application of nitrogen fertilizer. A crop in a less generous soil will need fertilizer sooner, and the question then arises of whether to put all the nitrogen on at that point or whether to put on perhaps one-fifth of the nitrogen then and the rest later. We see the question as being mainly one of risk of the fertilizer being washed out of the soil by rain. The earlier in the spring the fertilizer is applied, the more likely it is on average to be washed out of the soil. This is not only a waste of the farmer's money but an extra load on the environment, so the advantage seems to lie with holding back the main part of the application. Some farmers spread the risk by splitting the application into three parts but risk later applications being used inefficiently if the soil dries out. More than three seems unlikely because each one means fuel for the tractor and a few man-hours to be paid for.

We mentioned earlier the idea that the development of the crop could be manipulated to the farmer's advantage by using several applications of nitrogen timed to coincide with certain stages in the development of the crop. This is in the nature of fine-tuning relative to what we said above.

Some plant physiologists think that this can be done, while others disagree. The evidence for this happening with cereal crops seems equivocal. As soil scientists, we do not intend to rush into the fray.

Lastly, a word about applying nitrogen fertilizer in the autumn for winter crops. It used to be more or less taken for granted that autumn-sown crops needed a little nitrogen to help them through the rigours of winter, and special compound fertilizers were made for the purpose. Recent experiments have nearly all suggested that there is no need for autumn fertilizer. The Brimstone experiment described in Chapter 5 measures nitrate lost from the soil, and Goss *et al.* (1988) found that in most years applying autumn nitrogen to winter wheat simply increased the amount of nitrogen carried out of the soil as nitrate during the winter. Thirty kilograms of nitrogen as fertilizer in autumn simply meant 30 kg of extra nitrate-nitrogen in the drainage (Fig. 2.5). Nitrogen fertilizer is just not needed in autumn because the soil is fairly warm and getting moister, giving conditions in which the soil's microbes are well able to supply the nitrate that the crop needs. The same figure shows losses of nitrogen from fertilizer applied in spring to the same experiment and as you can see the losses are relatively small.

Fewer and fewer farmers now apply nitrogen in the autumn which is a good thing because such nitrogen might have contributed to the increase in the concentration of nitrate in natural waters. When we talked to farmers at an agricultural event last year most of them said, no, they did not use autumn nitrogen any more. Quite a few, however, said that they only used it on winter oilseed rape. 'Why rape?' we asked. 'Because of the pigeons', they said. It seems that pigeons, like aircraft pilots, do not like to land in heavy vegetation. A dose of nitrogen in the autumn can thicken the crop enough to make the pigeons cautious, but there is no evidence that it serves any other purpose. Fortunately, rape makes rapid use of nitrogen, and the autumn nitrogen that it receives may be less of a pollution problem than that given to winter wheat. The nitrate problem certainly has a great number of facets.

The wider picture

Most of this chapter has been concerned with arable crops grown under conditions found in the UK. Broadly similar conditions occur in other parts of Western Europe, except the Mediterranean countries, and much of what we have said should apply in these countries and also in parts of North America. Further afield, where the climate differs appreciably from ours, much of the detailed information that we have given may be inapplicable. We hope, however, that the general principles that we have laid down will

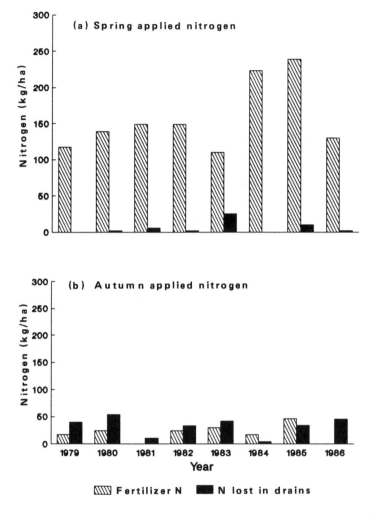

Fig. 2.5. Losses of nitrogen in water draining from the Brimstone experiment from (a) fertilizer applied in spring and (b) applied in autumn.

apply, because nitrogen has broadly the same functions in all crops. The aim should be to ensure that the crop is not held back for lack of nitrogen but that as little mineral as possible is in the soil when it is not needed: in other words to synchronize the supply of mineral nitrogen with the crop's requirements. Some areas and some crops will not have a nitrate problem. Very acid tropical soils bind nitrogen (Chapter 3) and these presumably do not have nitrate problems – except perhaps of deficiency. Wetland rice is grown in paddies under several centimetres of water in conditions in which nitrate will neither be produced nor survive. The crop absorbs its nitrogen in the ammonium form and urea is usually used to supply its nitrogen.

Not all of Britain is put down to arable cropping. About half the fertilizer that is sold in the UK is used on grassland. The grasses and the cereal crops are closely related in botanical terms and the main difference between them is the practical one that the farmer wants leafy growth from grass and grain from cereals. (A few farmers grow grasses to produce seed.) The emphasis on leafy growth means that more nitrogen fertilizer can be applied usefully to grassland than to cereal crops since grass mops up nitrate very effectively. About 400 kg of nitrogen can be applied safely to each hectare of grassland – provided the grass is not to be grazed by animals. This leads us to an interesting but rather complicated story that can be found in Chapter 8.

The way in which land is used can often be said to be 'intensive' or 'extensive'. Intensive land use means that the inputs to the system – time, money, effort, fertilizer, crop protection, etc. – are considerable, and they are made to ensure a substantial and profitable output of grain, sugar or whatever the crop yields. Extensive use of the land means that both the inputs and the returns are smaller. The production of wheat in Britain is intensive, at least partly because of the high cost of the land and high interest rates. In most of Australia and many other parts of the world the limitations imposed by the climate and the land make it an extensive activity. Animal production divides similarly. The production of meat, milk and dairy produce is highly intensive in The Netherlands, where land is at a premium, but the sheep dotting the hills in northern and western Britain reflect an extensive approach, as do ranch-managed stock in the drier parts of Australia and North America. Almost needless to say, the more intensive the system, the more nitrogen fertilizer it uses and the more nitrate is at risk.

The extreme of intensive agriculture is vegetable growing. Suitable land is expensive and the crops fetch high prices. In this situation nitrogen fertilizer is so cheap relative to other costs that the optimum is barely definable in economic terms and the best results are often obtained with amounts of nitrogen that are very large in relation to the quantity removed in the produce. Researchers have found vegetable soils to contain large amounts of nitrogen as nitrate – sometimes more than 500 kg ha^{-1}, about 10 times the amount left by winter wheat.

At the other extreme we find forestry. Those who invest in trees do not expect a rapid return on their money. Little, if any, nitrogen is supplied when trees are planted and they subsequently scavenge nitrogen very effectively from the soil (and also from the air). Water draining from forests usually contains little nitrate because most of the nitrogen is cycled from the soil to the trees and then back to the soil in leaf-fall in a steady cycle. Any problems that do occur usually come about when the trees are felled: the nitrate-scavenging part of the system has been taken out and quite substantial amounts of nitrate may drain from the system. Two problems have arisen that will return in other contexts: the soil is bare and an established cycle has been broken.

Chapter Three
The Physics and Chemistry of Nitrate Leakage

The first stage in dealing with any problem is understanding why it is occurring. To do so we need to examine the problem from the aspects of the three main disciplines of natural science: chemistry, physics and biology. This chapter deals with the chemistry and physics of the problem while the following one discusses the biology involved.

Some basic chemistry

The press often talk of nitrates – in the plural – while we usually refer to nitrate. It is true that there are a number of nitrates, sodium nitrate, potassium nitrate and calcium nitrate to name but three, but we are concerned with nitrate in water. When, for example, potassium nitrate dissolves in water it does not remain as potassium nitrate. It undergoes a process known as 'dissociation'. This means that the potassium and the nitrate become independent entities in the solution. So if potassium nitrate and calcium nitrate are dissolved in the same solution there is no way of telling whether a particular nitrate was previously associated with potassium or calcium. A laboratory shelf with bottles of potassium, sodium, calcium and magnesium nitrates can rightly be said to hold nitrates. If, however, you tip the contents into a large bucket of water and they all dissolve, the water has nitrate in it; it is quite irrelevant to say nitrates.

When potassium nitrate (KNO_3) dissociates, the K and the NO_3 both have an electrical charge; the K becomes K^+ with a positive charge while the NO_3 acquires a negative charge to become NO_3^-. These positively- and negatively-charged entities are known as *ions*. The soil solution has to contain the same number of positive and negative charges but this does not necessarily mean the same number of positively- and negatively-charged ions because some ions carry two or more charges. Calcium for example is Ca^{2+} while sulphate is SO_4^{2-}. Phosphate can have one, two or three charges according to how acid or alkaline the soil is. Nitrate always has a single negative charge. We realize that this basic chemistry will have been 'old hat' for many readers. Its inclusion was inspired at least in part by the experience of the advisory service manager from a fertilizer company[1] who went to give a talk at a meeting organized by an old-established agricultural merchants. He referred to ions and at the end of the talk he was approached by the patriarch of the merchants. 'Well young man, that was a really interesting talk. I had no idea that there was so much *iron* lying around in the soil'. We hope that the inclusion of the basic chemistry will help other non-specialists. The main point is that nitrate in solution is a negatively-charged ion and the significance of this should become apparent shortly.

The other ion that is important in this context is the ammonium ion, NH_4^+. As its name implies this ion is a close relative of ammonia NH_3, whose pungent smell may be familiar. Like nitrate ammonium exists as a single entity only in solution; the relation between ammonia, water and ammonium can be summarized as follows. Ammonium hydroxide is in the middle.

$$NH_3 + H_2O \rightleftharpoons NH_4OH \rightleftharpoons NH_4^+ + OH^- \tag{3.1}$$

The formation of the hydroxyl ion, OH^-, means that dissolving ammonia in water gives an alkaline solution. The double arrows mean that the change can go in either direction and one consequence of this is that some ammonium applied as fertilizer can become ammonia, which is a gas and can be lost to the air. This happens mainly in soils that are already slightly alkaline and during weather that causes the soil to dry, that is, warm and windy weather. The main point about the chemistry of ammonium in this context is that it carries a positive charge when dissolved in water.

We need here to make a brief foray into the next, biological, chapter and say that ammonium is converted to nitrate by microbes in the soil. This process, known as nitrification, happens readily in most soils, particularly when the soil is moist and reasonably warm. Obviously, the physical and chemical properties of ammonium remain relevant only until this happens.

Is nitrate chemically bound to the soil?

The nitrate problem happens because water, usually rain water, carries nitrate with it when it passes through and out of the soil. Anything that holds the nitrate back helps to ameliorate the problem. It would be useful, for example, if nitrates, or even one particular nitrate, did not dissolve readily in water so that a proportion was always left in the soil in solid

---

[1]We are grateful to Paul Thompson for this anecdote.

form. This is not so, unfortunately; nitrates are conspicuously soluble. Calcium is usually the main ion found in the water held in British soils and calcium nitrate is so soluble that 1 kg of it will dissolve in a litre of water – if you are patient and keep stirring the mixture. Indeed calcium nitrate has so strong an affinity for water that if you put some out in an open container it will pick up water from the air and dissolve in it.

It would also help if nitrate adhered to the surfaces of the particles in the soil, but this does not happen in most soils in the temperate zone. Finely-divided matter often does absorb other substances from solution. Charcoal for example, is useful for removing some impurities from water. The finest particles in soils are the clays and these too can be useful for adsorbing some substances – but not necessarily nitrate. The problem is that clays readily accumulate electrical charges, and whether the overall charge is positive or negative depends on the pH, which is the measure used to quantify the acidity or alkalinity of the soil. Most temperate agricultural soils are given lime to ensure that they do not become acid, and their pH values are usually in the range 5.5–8, which means slightly acid to slightly alkaline, pH 7 being neutral. In these conditions the clay particles carry an overall negative charge. This means that positively-charged ions such as potassium, calcium, magnesium and ammonium are attracted to the surface of the clay, but negatively-charged ions such as nitrate and chloride are repelled (Fig. 3.1). (Like repels like with electrical charges.) So, far from being held in safety by the clay particles nitrate is forced to keep its distance from them and is made more vulnerable to being washed out.

Not all soils have a negative charge. In the wet tropics the weathering processes that produce and gradually modify soils tend to proceed faster because of the heat and the large amount of water passing through soils. A consequence is that soils in the humid tropics tend to be acidic and often have an overall positive charge on the clay. In many such soils, therefore, the nitrate is attracted to the clay and experiments show that the nitrate is to some extent held back against being washed out (e.g. Wong *et al.*, 1987).

Ammonia and ammonium

In whatever we have said about nitrate so far the reverse is true when we turn to ammonium, because we are considering a positively-charged ion rather than a negatively-charged one. Ammonium is strongly attracted to negatively-charged clay particles and is rarely washed out of a non-acid soil in any quantity, unless the soil is a very sandy one with little clay. Thus we have no 'ammonium problem' in our natural waters. If we could keep the ammonium as ammonium longer by discouraging the microbes in the soil from converting it to nitrate, we would have less of a nitrate problem. There are various chemicals that can be used to slow down the nitrification

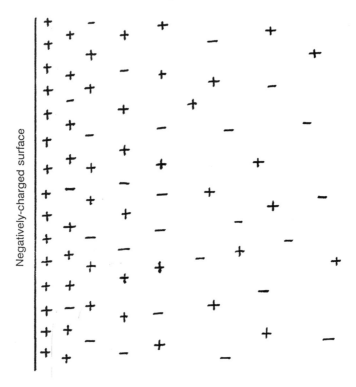

Fig. 3.1. A simplified diagram of the distribution of positively- and negatively-charged ions near a negatively-charged surface.

of ammonium by the microbes. One of them, curiously enough, is ammonia. Ammonia can be liquified under pressure and then used as a nitrogen fertilizer. It is injected under pressure into slits in the soil using a special machine. Once in the soil the ammonia reacts with water in the soil as shown in Equation (3.1) to produce ammonium. Either the large local-ized concentration of ammonia or the intense alkalinity has an inhibitory effect on the microbes that only wears off slowly. Another simple inhibitory chemical is carbon disulphide but this is a toxic substance with a vile smell so it cannot easily be used as an inhibitor. A better approach is to use a chemical such as sodium or ammonium trithiocarbamate that releases carbon disulphide as it decomposes in the soil. The chemical companies have developed various other inhibitors of nitrification the best known of which are 'N-serve' and dicyandiamide (DCD).

The effectiveness of these nitrification inhibitors depends to quite a large extent on where they are used. In large areas of Northern USA and Canada the soil remains frozen for much of the winter. Fertilizers containing ammonium or ammonia can be applied in the autumn with inhibitors that keep the ammonium as ammonium until the soil freezes.

Once this happens the cold inhibits the microbes and the fertilizer remains safe until spring. The thaw means that a large amount of water passes through the soil, but so long as the fertilizer has remained as ammonium little is lost. In much of western Europe the soil freezes only rarely and is warm enough for microbes to function, albeit in a limited way, for much of the winter. There would usually be no point in applying ammonium fertilizers, even with inhibitors, in autumn for the crop to use in spring. What might be more profitable, however, would be to use inhibitors to keep indigenous ammonium as ammonium, and thereby lessen the amount of nitrate in the soil for as long as possible in autumn.

Nitrogen fertilizers

Nitrogen fertilizer usually contains nitrogen in one of three forms, ammonium, nitrate or urea. Of these the commonest, on a world basis, is urea, which in chemist's terminology is an organic compound, $CO(NH_2)_2$. It occurs naturally in urine, but is manufactured by a synthetic process. Its main advantage over other nitrogen fertilizers is that 46% of its weight is nitrogen, more than any other form. This becomes more relevant the further you have to transport the fertilizer. In soil urea is rapidly broken down by an enzyme, urease, to ammonium. British farmers use ammonium nitrate to quite a large extent, mainly because of a policy decision by the then largest manufacturer, and mono- and di-ammonium phosphates are incorporated in compound fertilizers that contain phosphate as well as nitrogen (and usually potassium as well). Calcium nitrate is used in some parts of Europe. The activities of soil microbes usually mean that whatever form of fertilizer is used the end product in the soil is nitrate. It does not matter where the nitrate has come from, from calcium nitrate, from an ammonium fertilizer, from urea (synthetic or in cattle urine), by deposition from the air or from organic nitrogen broken down in the soil by microbes, the nitrate is exactly the same. The nitrogen atom and three oxygen atoms will be arranged in a symmetrical planar structure that can be described with great exactitude by a chemist. There is no way in which a chemist, or anyone else, could tell you what kind of farming system has produced a particular nitrate ion.

The root of the nitrate problem

The underlying principle of the nitrate problem, like the structure of the nitrate ion, is the same regardless of the origin of the nitrate. It can be summarized in the form of an 'equation'

$$availability = vulnerability$$

Any nitrogen in the soil that is *available* to crops is almost always in one of two forms: ammonium, which is rapidly converted to nitrate by soil microbes, or nitrate itself. We have seen that nitrate dissolves freely in water and is also not held on soil particles in non-acid soils and is therefore *vulnerable* to being washed out of the soil by rain water (or irrigation water) passing through the soil.

The surest way of decreasing losses of anything is to have as little as possible to lose. This means having as little nitrate as possible in the soil. When crops are growing fast they need a generous supply of nitrate, but once they have ceased to grow and to absorb nitrate from the soil we need to make sure that there is as little nitrate as possible there. Any nitrate that is in the soil is vulnerable to being washed out and is there at the wrong time. We can therefore define the root of the nitrate problem as *untimely nitrate*.

Whether or not untimely nitrate becomes nitrate pollution depends on what happens to rain or irrigation water falling on the soil. This in turn depends largely on the physical behaviour of water in the soil, so the remainder of this chapter is devoted to a brief account of how the physics of the soil determines the downward movement of water and nitrate. Before we turn to soil physics, however, we need to mention a phenomenon that straddles the realms of both chemistry and physics – diffusion.

Diffusion

Water poured into a beaker or similar vessel flows to achieve a level surface. Put another way, liquids distribute themselves in any vessel so that their potential energy, their potential to flow downhill, is the same at all points. If, rather than pouring water into an open vessel, we introduce steam or another gas, into a closed vessel something similar happens. The gas distributes itself such that its pressure is the same at all points; pressure is another form of potential energy. If there is a mixture of gases, each exerts its own 'partial pressure' and distributes itself such that its partial pressure is the same throughout the vessel enclosing it. If the partial pressure of a gas is changed in any part of a system, there will be a gradient of partial pressure and a gas will flow in such a way as to equalize its partial pressure throughout the system. This process is known as *diffusion*; the flow is *diffusive flow*. One important example of this process happens in the soil when living organisms consume oxygen and give out carbon dioxide. Oxygen diffuses into the soil in response to the gradient in its partial pressure, while carbon dioxide diffuses out of the soil in the same manner but in the opposite direction. We shall return to the diffusion of oxygen in Chapter 4.

A solute in solution behaves in several important ways like a gas in a closed vessel, because it has potential energy, its chemical potential, which is related to the partial pressure of the gas. In a similar way to gases a solute distributes itself so that its concentration is the same throughout the solution. Changing the concentration in any part of the system causes a gradient in concentration down which solute will move such as to equalize the concentrations. This process is another form of diffusion, and the diffusion of solutes in water in the soil is important for at least two reasons. One is in plant nutrition: a root that removes nitrate or any other plant nutrient from the soil immediately around it creates a concentration gradient down which further supplies of the nutrient move to replace what has been taken by the root. Diffusion is most important in this way for nutrients such as potassium which are in the soil solution at small concentrations rather than nitrate which is often in the soil solution in much larger concentrations. It is also relevant to the nitrate problem, as will be seen later in this chapter, because it is one of the main factors that determines whether or not nitrate is protected from leaching.

Some basic soil physics

If you dig out a block of soil without disturbing its structure, dry it and then weigh it, you will find that the ratio of the mass to the volume, that is the bulk density, is about 1.3 g cm^{-3}. The exact value will depend on the type of soil. If you then crush it and drive out all the air, the density will be about 2.6 g cm^{-3}. Solid matter therefore occupies only about half the volume of the soil. Air and water make up the rest, and there is apparently little to stop water progressing steadily downwards through the soil and carrying nitrate with it. In some types of soil this is true, but in others the water moves downwards with difficulty or not at all. Much depends on the sizes of the soil particles and their structural arrangement.

The sizes of particles in soils are so important that they have an international classification. Four fractions are defined (Table 3.1).

The sand fractions obviously feel gritty when touched, but at the other end of the scale the clay particles are so fine that if you wet them thoroughly and then rub them between your fingers they feel almost soapy. There are soils that are almost all sand, all silt or all clay but most soils contain a mixture of all three. It is very common to find that the texture of the soil, that is the relative proportions of sand, silt and clay, changes as you dig deeper. You may find that it changes as you go from one part of the field to another. This 'spatial variability' is one of the complicating factors in any attempt to measure or describe the washing out of nitrate from soils.

When we talk about 'clay', we should really say 'clays' because we are talking about a family of materials that have the same general nature but

Table 3.1. The International Soil Science Society's classification of particle sizes.

Fraction	Particle diameter (μm)
Coarse sand	2.0–0.2
Fine sand	0.2–0.02
Silt	0.02–0.002
Clay	<0.002

Larger particles are classified, in order of increasing size, as gravel, cobbles, stones and boulders. These are not usually considered to be soil material.

show differences in the way that their constituents, aluminium, silicon and oxygen are arranged. Clays are remarkable materials for the way in which the particles adhere to each other. Were it not for this property there would be no such craft as pottery and our every day life would be rather different. These adhesive properties are also very important in soils. The clay acts as a cement that holds other particles together and thereby has a strong influence on the structure of the soil. One consequence of this can be a layer of clay in the subsoil (that is, the soil beneath the zone affected by cultivation) that is virtually impenetrable to water, so that the soil has to have some form of man-made drainage. Many clays also have the capacity to imbibe water so that they swell when wetted and shrink when dried. This means that if the roots of growing crops manage gradually to draw water out of a clay subsoil it eventually shrinks and cracks so that water and roots can penetrate it. Such subsoils often tend to shrink and crack during summer and then swell and close during winter. Sometimes the cracks become permanent. The Dutch Polders, areas of land reclaimed from the sea, have very heavy clay in places. During the reclamation process the soil is gradually dried out so that cracks form and allow water to drain. This process, known as soil 'ripening', is fascinating to see and the appearance of the soil at the end of the process is quite different from that at the start. Cracking clay soils known as vertisols are important agriculturally in many tropical areas.

The cementing properties of clay are also important in the topsoil, the part which is cultivated. Ploughing, harrowing and other cultivations fragment the soil, as do wetting and drying which cause swelling and then shrinking and cracking. On top of this the topsoil may freeze; this causes the water it contains to expand and shatter it further. The topsoil also contains another, rather different, cementing material – the soil organic matter. This comprises matter from generations of plant residues that has been decomposed by microbes in the soil (Chapter 4) and converted to a form in which it is able to coat individual clay particles and also the fragments produced by the various shattering processes. In doing so it helps to cement the fragments together and also stabilizes them against being broken into smaller and smaller fragments.

The outcome of this diversity of activity is what we call the soil structure. The key feature of the structure in the topsoil are the 'aggregates' or 'crumbs', which are the cemented and stabilized fragments whose genesis was described in the previous paragraph. All agricultural soils are aggregated to some extent, but the degree of aggregation depends on the texture of the soil. In general, the more clay a soil contains the more aggregated it will be. This is partly because of the cementing effect of the clay but it also reflects the fact that the clay and organic matter tend to have a mutually stabilizing effect, so that more clay is likely to mean more stabilization by organic matter as well.

Farmers are concerned to get a good 'seed-bed' for their crops. This means the best possible environment for the germination of the seed and the early growth of the young plants. For this the aggregates in the soil needs to be 0.5–2 mm in diameter, that is, roughly the same size as the coarse sand particles. Aggregation is therefore less necessary in soils containing a lot of sand and it is also less likely because the more sand a soil contains the less clay it will contain, unless the difference is made up by the third particle fraction, silt. The larger sand particles are not easily bound into aggregates and tend to break them apart if incorporated.

We have seen how the structure of the topsoil depends on cultivation, wetting and drying, freezing and organic matter. Of these factors only wetting and drying affect the subsoil to any extent, and then only in clay subsoils (see above). In clay subsoils again, layers that are impenetrable to water and the roots of crops may be subjected to an operation sometimes described as 'subsoiling' or 'soil busting' in which an implement is torn through the layer to make pathways. This process uses much energy and is only undertaken when definitely necessary. Structure in the subsoil therefore depends greatly on the inherent properties of what is there naturally. The least difference between topsoil and subsoil is probably found in sandy soils where the difference consists mainly in organic matter in the topsoil. Silt soils will also show little difference where the subsoil is silt; silt, however, is quite often found deposited on other materials. The greatest difference usually comes with clay soils. Here the topsoil will consist of aggregates of various sizes ranging from 0.5 mm upwards and probably including some larger lumps in which the aggregates are bound together. A clay subsoil on the other hand will be a much more massive structure whose units are defined by lines where the structure has split when drying and shrinking occurred and which are to be measured in centimetres or even tens of centimetres rather than millimetres.

The movement of water in the soil

Water may exist as a gas (vapour or steam) a liquid or a solid (ice). All three phases may be found in the soil, but we are concerned with water as a liquid; the other phases do not wash nitrate out. Water may fill up all the pores in the soil, in which case the soil is said to be *saturated,* but a substantial proportion of the pores usually contains air as well; the soil is then *unsaturated.* The fact that the soil is unsaturated does not mean that the water cannot move; water flows in both saturated and unsaturated soils. In fact having the soil saturated is bad, because air cannot flow in the soil and this restricts the growth of plant roots and alters the behaviour of microbes (p. 50).

Water flows in response to a force of some kind. The force may be that resulting from gravity, which makes it flow downwards through soil or it may be in a difference in pressure. Water in the soil, however, is subject to subtle pressures, particularly that in very fine pores. The surface tension of the water in soil pores causes a curved meniscus (Fig. 3.2) across which there is a difference in pressure. This is similar in principle to the minute pressure difference between the inside and outside of a soap bubble that keeps the bubble inflated. The sharper the curvature the greater the pressure difference is; small bubbles are more durable than large ones. Similarly the finer the pore is, the sharper the curvature is and the greater the pressure difference across the meniscus. Figure 3.2 shows how this pressure difference forces water into a pore. The water beneath the

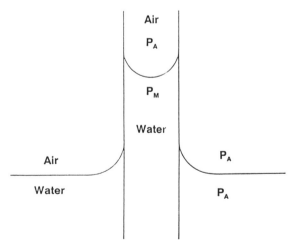

Fig. 3.2. Capillarity. Because of the curvature of the meniscus P_M the pressure below the meniscus is less than P_A, the atmospheric pressure. The surrounding water is at pressure P_A, so water is forced into the capillary by the resulting pressure difference.

meniscus is at a pressure less than that at the atmosphere and is forced into the pore by water at atmospheric pressure. This phenomenon is often known as capillarity; the pores involved have about the same diameter as a hair, and *capilla* is Latin for 'hair'. Forces of this kind can draw water upwards when the soil at the surface is drier than that lower down. They also move water to plant roots.

Fine pores are also significant in another way. Any pore transmits water such that the flow is proportional to the fourth power of its radius. So the flow in a pore 0.1 mm in diameter is 10000 times smaller than that in a 1 mm pore if both flows result from the same force. With the nitrate problem we are mainly concerned with downward movement of water and nitrate. Fine pores help to hold back this flow, and they do so in two ways. One is by the slowness of flow through them, and the other arises from the fact that they are located mainly in the aggregates in the topsoil, and in some soils, in the structural units of the subsoil. Water always takes the easiest path down through the soil and so tends to go round aggregates rather than through them (Fig. 3.3). The overall effect is often simplified to the concept that there are two categories of water in the soil, mobile and immobile; obviously the former is responsible for practically all the water flow. This phenomenon is one reason why aggregation has attracted interest from soil scientists. The degree of aggregation depends on the texture of the soil (see above). Clay subsoils also usually have the largest structural units. These mobile and immobile categories of water are

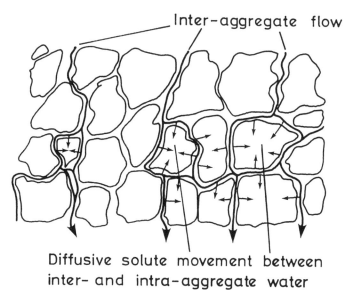

Fig. 3.3. Solute and water movement in an aggregated soil.

therefore most clearly defined in clay soils and least so in sandy soils. Water usually moves freely through sandy soils, much more so than in clay soils, but this is not to say that it moves uniformly. Some Belgian experimenters (De Smedt *et al.*, 1986) found evidence of immobile water even in a carefully packed column of pure sand in a laboratory and they would surely have found more evidence of it in a sandy soil in the field. Silt particles are intermediate in size between those of clay and sand, and silt soils tend to be intermediate between clay and sandy soils in their behaviour. They allow water to move without much difficulty, as do sandy soils, but they have the capacity to hold much more water than sandy soils, so that any particular spell of rain causes a smaller downward movement.

Soil structure and water movement: the implications for nitrate losses

The simplest way in which rain or irrigation water can wash nitrate out of the soil is described as 'piston flow' or 'piston displacement'. The concept is that the water entering the soil moves downwards, pushing in front of it the water containing nitrate that was previously in the same body of soil. Ten millimetres of rain displaces an equivalent volume of soil water. However, because the soil has a solid phase that does not move, the distance moved in the liquid phase is more than 10 mm. If, for example, only half the soil is occupied by water and the rest by solid matter the distance that the water is displaced is $10/0.5 = 20$ mm. This would only happen if the soil was completely uniform, which means practically never. In reality displacing water would move through different pores at different speeds so that the 'front' of displacing water would become spread out and 'dispersion' would occur. This would encourage the physical mingling of the displacing and displaced water and thereby tend to equalize the concentrations of nitrate between them. This mingling would blur the front and further blurring would be caused by diffusion of the nitrate within the moving water. Clearly the front is very far from being an even surface like the face of a piston, but dispersion and diffusion are only the beginning of the complications and may not even be the dominant effects.

We saw above how in aggregated soils in particular there tend to be two categories of water, mobile and (relatively) immobile, with the latter mainly in the fine pores in aggregates. Clearly nitrate that is held in water in fine pores in aggregates is safe while it remains there, because percolating rain will pass round, rather than through the aggregates (Fig. 3.3). This safety is temporary, however, because the nitrate will tend to move out of the aggregates by diffusion (see above) in response to the difference in nitrate concentration between the water inside and outside them (Fig. 3.3). But if nitrogen fertilizer has recently been applied to the soil surface, the mobile water may have a greater nitrate concentration than that in the fine pores so that the diffusion will be inwards rather than outwards and will protect the nitrate.

In the flow described above the mobile water interacts with the immobile water in fine pores in the sense that nitrate can diffuse from one to the other. Also, if the fine pores contain no water, their capillarity may force mobile water and nitrate into them. We need to consider too a category of water movement, often known as 'channelling' or 'by-pass' flow (Beven, 1981), in which water moves rapidly down through the soil without any appreciable interaction with the resident water or nitrate. Such flow happens through very large and continuous pores or cracks, commonly when more rain falls than can be absorbed by, and flow down through, the main body of the soil. In terms of keeping nitrate in the soil such flow can be beneficial because it carries water through the soil without displacing very much nitrate. If, however, it happens shortly after fertilizer nitrogen has been applied, the nitrate from the fertilizer can be carried rapidly through the soil without interacting much with the main body of the soil and therefore largely lost.

Assessing soils for risk of nitrate loss

The more water a soil can hold, the more has to be displaced for losses of nitrate to occur, so the *capacity* of the soil to hold water is important for restraining nitrate losses. This remains true in soils with clearly defined mobile and immobile categories of water because this capacity will determine the size of the immobile category in which nitrate remains temporarily safe from loss. The other main factor is the *rate* at which water moves through the soil. The faster it moves the greater the risk of loss.

The usual capacity factor is the soil's volumetric moisture content, commonly given the symbol θ_v. Because pores of different sizes are able to exert differing forces on water θ_v has to be defined for a given suction. One relevant suction is that at which water ceases to drain from the soil, but this suction differs between soils. The usual rate factor is the soil's hydraulic conductivity, K. The main problem with K is that it can easily differ by a factor of 100 between two points 1 m apart in the soil making it very difficult to estimate for a large area. Another problem is that K depends strongly on the amount of water held in the soil. It can often be measured without too much difficulty in saturated soil, but it is usually the value in unsaturated soil that we need to know and this is far more difficult. Both θ_v and K depend on the texture of the soil and this is often the most useful guide to the leakiness of a soil. All soils in England and Wales have been surveyed and classified by the Soil Survey and Land Research Centre who have produced maps that show how the risk of nitrate loss varies from place to place.

Chapter Four
The Biology of Nitrate Leakage

Life in the soil

'I don't see how you can get interested in dirt', said a friend when one of us decided to specialize in soil science. The soil, however, is very much more than dirt; it gives us life. One of the main things that makes soil more than just dirt is also central to the understanding of the nitrate problem, and this is the organic matter in the soil. Most of this organic matter is dead. It comprises the remains of plants, mostly long dead, that grew in the soil. This part is often called 'humus' and it is the humus that usually makes the topsoil darker than the subsoil. The other part of the organic matter is very much alive, being made up of the soil's micro-organisms, often known collectively as the 'microbial biomass' or sometimes just 'biomass'. The humus and the biomass depend critically on each other, and between them they govern the fate of most forms of nitrogen entering the soil. By the same token they have a major effect on the losses of nitrate and other nitrogenous compounds from soil.

Humus – dead organic matter

Organic matter in the soil is the third largest repository of nitrogen in the planet, coming after the atmosphere and the sea (Table 4.1). Even a soil that has long been cultivated and therefore has a relatively small organic matter content will contain about 2000–3000 kg (2–3 t) N ha^{-1}, nearly all in organic forms and mostly in the plough layer (the top 25 cm). More typical arable soils will contain 3–5 t N ha^{-1} and a peat soil will contain much more. The average application of nitrogen to winter wheat in the UK, 180 kg ha^{-1}, seems modest by comparison, but comparing the two is of limited value because the organic nitrogen in the soil is relatively inert and cannot be leached, whereas fertilizer nitrogen can. Although the organic matter derives from plant debris, it rarely remains recognizable as such for

Table 4.1. Global distribution of nitrogen (after Jenkinson, 1990).

Nitrogen	Tonnes
The atmosphere	3.9×10^{15}
Soil (non-living)	1.5×10^{11}
Microbes in soil	6×10^{9}
Plants	1.5×10^{10}
Animals (land)	2×10^{8}
People	1×10^{7}
Sea (various)	2.4×10^{13}

long. This is because it has been 'processed' by the soil's micro-organisms – or biomass. One eventual consequence of this processing is the production of nitrate, so we need to take a close look at the organisms responsible.

Biomass – active organic matter

The soil's organic matter provides both a home and a source of food for a large and varied population of micro-organisms that includes many species of bacteria, fungi, protozoa and algae. Sometimes it is convenient to treat these collectively as the soil microbial biomass but for some purposes we need to divide and classify the population and consider the activities of a single species of organism. A few species of bacteria have been studied in great detail but many other organisms remain shrouded in mystery because microbiologists do not have the means to study them; many of the techniques that they use in other spheres of science are simply not appropriate to soils. A medical microbiologist for example, can count the bacteria in a blood sample by diluting the sample and placing a subsample on a film of agar on a petri dish. The agar, a kind of jelly, contains a plentiful supply of nutrients and within a few days each bacterium grows into a colony that can easily be seen and counted. However, questions arise when this technique is applied to soil bacteria. It works, in that colonies of bacteria grow and can be counted, but the results do not compare well with those from other methods. A typical result could be that the agar plate showed 10 million bacteria per gram of soil. This seems a huge number, but viewing the soil directly with a microscope after it is treated with a selective stain will show 10 or even 100 times more bacteria. The reason for the difference is that the agar medium is far richer in microbial nutrients than the soil. Soil bacteria tend to be hardy species well adapted to life under near-starvation conditions and for them the agar is simply not an appropriate medium. Many of the other organisms are far more difficult than the bacteria to isolate and count.

Nitrogen in the soil's food chain

Plant residues are normally the main form of organic matter entering the soil, but they are occasionally supplemented by dead animals. These residues are sometimes attacked first by fungi; a web of white fungal hyphae can often be seen on dead leaves lying on the soil or on wheat straw ploughed into it. More often, however, the first line of attack comes from small soil animals such as earthworms, millipedes, springtails or, in tropical areas, termites. These break up plant and animal residues and mix them into soil, thereby bringing them into more intimate contact with the soil's microbial population, which plays the main part in the break-down of the materials into humus. Each of the organisms involved in breaking down the residues retains some of the nitrogen in its own cells to manufacture essential molecules such as proteins and nucleic acids, and the rest is returned to the humus – the non-living part of the soil's organic matter. The microbes in the soil that break down humus eke out a precarious existence under the near-starvation conditions mentioned above, because most of the humus is not easily decomposed. They need carbon from the humus to obtain energy and for building cells and nitrogen for the essential molecules mentioned above. The carbon and nitrogen that microbes do not use for these purposes are either returned in a chemically altered form to the humus or released in the soil as simple non-organic waste products; carbon dioxide from the carbon and ammonium from the nitrogen. This process is normally called *mineralization,* and it also occurs when cells die. The carbon dioxide diffuses out of the soil but the ammonium remains, often held on the surface of a clay (Chapter 3).

Some soils are very acid or permanently waterlogged. In such soils ammonium is the final product of mineralization, but in most of the world's soils ammonium is oxidized by microbes, first to nitrite and then to nitrate, such that:

$$\text{Organic N} \rightarrow NH_4^+ \rightarrow NO_2^- \rightarrow NO_3^-$$

ammonium nitrite   nitrate

Nitrite very rarely accumulates in soils, which is fortunate because it is toxic to most soil organisms. Thus the final product of mineralization is usually nitrate. The first step of the overall process of mineralization, the conversion of organic nitrogen to ammonium. is called ammonification, while the conversion of ammonium to nitrate is called nitrification. A wide variety of fungi and bacteria can bring about ammonification, but nitrification is a more specialized process involving a small number of species of bacteria (for example of the genera *Nitrosomonas* and *Nitrobacter*). These are, however, to be found in almost all soils.

In one way it is unfortunate that nitrifying bacteria and nitrification are so ubiquitous in soils. Most plants can use ammonium as a source of nitrogen as well as nitrate (although there are some exceptions), and from an environmental viewpoint ammonium has a great advantage over nitrate. It has a positive charge and, as we saw in Chapter 3, it tends to be held on the negatively-charged clay surfaces, so that it is much less easily washed out of soil than nitrate. This comment does not, of course apply to tropical soils with positively-charged surfaces.

Predators and prey

Tennyson's phrase 'nature red in tooth and claw' applies almost as well in the soil as it does in the African bush. Predation of one species on another plays an important part in speeding up the process of mineralization. Some species are the 'vultures' of the soil, feeding on the dead bodies of other organisms, while others are more akin to hunters; protozoa for example, prey on bacteria. In each case the predator ingests and uses some of the nitrogen from the prey and excretes the rest as ammonium, which is usually nitrified. The speeding up of mineralization has been demonstrated in 'microcosms' – a rather grand name for test-tubes containing soil. If protozoa are eliminated in one set of microcosms but allowed to remain in another, the microcosms with the protozoa are found to contain much more nitrate than those without. Similar effects can be seen with earthworms and other soil animals.

The nitrogen switchback

The process of mineralization has been described so far as if the conversion of organic nitrogen to ammonium and nitrate was a one-way street. There is, however, a parallel street running in the opposite direction. This is the process called *immobilization*, in which soil organisms take ammonium and nitrate from the soil and convert them into organic forms of nitrogen. The reason for immobilization can be seen by considering the fate of two contrasting materials in the soil, dried blood and glucose.

Dried blood is a waste product from slaughterhouses that is used as an organic fertilizer. It is very rich in nitrogen containing about 10%, so when the soil's micro-organisms have obtained what they need from the dried blood they excrete substantial quantities of ammonium. Nitrate is therefore produced very rapidly when dried blood is applied to the soil. Glucose, on the other hand, contains no nitrogen at all. The microbes can use its carbon as an energy source, but if they wish to grow and divide they need nitrogen. If the surrounding soil contains ammonium or nitrate they consume it and incorporate it into organic matter. The ammonium and nitrate are thus immobilized so far as availability to plants is concerned – and also with respect to leaching. Adding glucose to soil in the laboratory can cause all the ammonium and nitrate to disappear within a few days, but once the glucose has been used up mineralization re-establishes itself as the dominant process.

Most organic material that enters the soil falls, with respect to its nitrogen content, between the extremes of dried blood and glucose. Whether the material causes mineralization or immobilization to predominate depends on the ratio of carbon to nitrogen it contains. As a rule-of-thumb, materials with a carbon to nitrogen ratio of less than about 25 tend to mineralize nitrogen as soon as they start to decompose. Those with a ratio appreciably greater than 25 tend to immobilize ammonium and nitrate for a while before mineralization re-asserts itself (see, for example, the effect of wheat straw on immobilization in Fig. 4.1). The carbon to nitrogen ratio is a useful guide to the probable fate of organic matter entering the soil but it is not infallible. The chemical or physical nature of the material plays a part in deciding which process predominates. In one laboratory experiment, for example, the roots of some clover plants caused immobilization for a short period of time even though they were rich in nitrogen (H. Kirchmann, personal communication). This could have been because the parts of the roots that decomposed first had a higher ratio of carbon to nitrogen than the roots as a whole.

It is quite usual for some of the organisms in a soil to be mineralizing nitrogen while others are immobilizing it simultaneously. Which group predominates depends on the types and amounts of organic matter that have entered the soil and on the relative contributions to the whole process of decomposition of the incoming fresh material and the humus already there. When mineralization is the dominant process it is relevant to think in terms of 'net mineralization'.

The destruction of nitrate by microbes – denitrification

The soil biomass not only produces nitrate; it also destroys it. Thirty-five years ago the distinguished American soil scientist F.E. Allison wrote a noted review paper entitled 'The enigma of soil nitrogen balance sheets.' The enigma lay in the many experiments in which attempting to balance inputs and outputs of nitrogen to and from the soil showed losses that could not be accounted for by leaching alone. In this review Allison (1955) drew attention to the importance of denitrification and other processes that form nitrogenous gases. Although chemical denitrification, in the form of

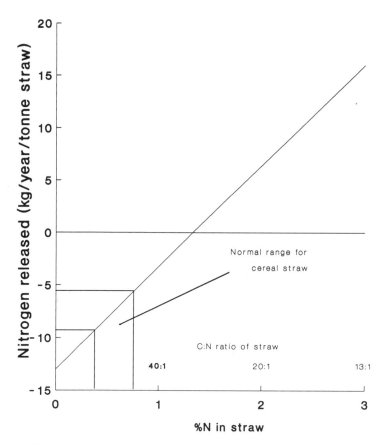

Fig. 4.1. Relationship between %N (C : N) in crop residues and the amount of mineral nitrogen released per tonne of residue incorporated (taken from Jenkinson, 1985).

spontaneous decomposition of ammonium nitrate, is possible when the soil is totally dry, most denitrification occurs when the soil is definitely wet, and this form of the process is brought about by microbes in the soil. The reason for including denitrification in a book about the nitrate problem is simply that any nitrate that is denitrified is not leached and therefore does not contribute to the nitrate problem. This does not mean, however, that denitrification is entirely beneficial in an environmental context as will be shown shortly.

The reason for denitrification lies in the need of the soil's microbial population for oxygen. Microbes use oxygen when they decompose organic matter in the soil and produce carbon dioxide, as we saw earlier. This is a form of respiration in which electrons are passed along a chain of intermediates to the oxygen and during which the microbes acquire energy. If oxygen is absent or present in very small concentrations the microbes can only obtain energy if they can find alternative acceptors of electrons, and nitrate is used to produce carbon dioxide in place of the oxygen from the air. Since denitrification is effected by oxygen-starved microbes it is clearly to be classified as an anaerobic process: nitrogen gas, N_2, or nitrous oxide, N_2O, is produced.

The soil usually becomes anaerobic simply because it is wet. As it becomes wetter, more and more pores are filled with water so that air cannot diffuse through them as quickly as when dry. Oxygen diffuses about 10000 times more slowly through water than through air. What this often means is that as the microbes use up oxygen, oxygen-deficient zones develop at the centres of the larger aggregates (Currie, 1961; Smith, 1980), but in extreme cases the entire soil can become waterlogged and therefore anaerobic. Wetness, however, is not the only factor involved; the microbes' demand for oxygen develops because there is material that they can decompose. A substantial fragment of easily decomposed plant material can lead to a very large demand from the microbes for oxygen that cannot be met by diffusion. The fragment then becomes the centre of intense denitrification, described as a 'hot spot' (Parkin, 1987). The existence of these hot spots means that the intensity of denitrification varies enormously within even a relatively small volume of soil, and this variability is one of the factors that makes denitrification extremely difficult to measure, particularly in experiments in the field. Another result of these hot spots is that nitrate can be produced by nitrification at one point and destroyed only a short distance away, an aerobic and anaerobic process occurring almost side by side.

The *potential* for denitrification to occur in soils is very great. Water-logged soils kept warm in the laboratory (at 25°C say) and supplied with plenty of easily-decomposable organic matter will rapidly use up the supply of oxygen, and the resulting denitrification may destroy nitrate at a rate equivalent to 30 kg N ha^{-1} day $^{-1}$. Rates of denitrification in soils in the field, however, are more likely to be of the order of 3 kg N ha^{-1} day^{-1}. Denitrification will also probably occur in short bursts following rain or irrigation, rather than continuously (e.g. Jarvis *et al.*, 1991). The rate of denitrification also depends fairly obviously on the supply of nitrate and is likely to be greatest when nitrogen fertilizer has recently been applied.

We have seen that denitrification is very variable in both space and time, and this makes it very difficult to measure in field experiments. Considerable research effort is being directed towards improving the methods of measurement.

When the emphasis in farming was simply on production, denitrification was clearly undesirable because it represented the loss of a valuable plant nutrient that would have to be replaced. With increasing concern about nitrate losses to natural waters, however, denitrification assumed a new and more favourable role, that of preventing nitrate losses into surface waters and aquifers. The desirability of denitrification depends, however, on the gas that is formed. Nitrogen gas, N_2, is no problem because it makes up 78% of the atmosphere in any case, but nitrous oxide is implicated in two major environmental problems, climate change and the depletion of the ozone layer.

Nitrous oxide, like carbon dioxide, absorbs infra-red radiation from the sun and converts it to heat, thereby contributing to global warming. Nitrous oxide is present in the atmosphere at a far smaller concentration than carbon dioxide. It is not, however, a far lesser problem, because it is 180 times more effective than carbon dioxide as a 'greenhouse gas' and is currently estimated to account for 5–10% in the potential increase in global warming.

Ozone, O_3, is a relatively rare form of oxygen found in a zone of the upper atmosphere. There it acts as a protective shield, filtering out some of the ultraviolet radiation from the sun that would otherwise contribute to skin cancer. When nitrous oxide, N_2O, enters the upper atmosphere it undergoes chemical changes that form nitric oxide, NO, and this gas is involved in reactions that destroy ozone.

A substantial proportion of the atmospheric nitrous oxide that causes these problems comes from the soil. On a global scale the amount of nitrous oxide emitted from soils is twice the amount produced by burning fossil fuels and four times the amount evolved from the ocean (Jenkinson, 1990). Any agricultural practice that increases the production of nitrous oxide must clearly be avoided, and we need to ensure, as far as possible, that any denitrification that does occur produces nitrogen gas, rather than nitrous oxide.

The ratio of nitrous oxide to nitrogen in the gases formed during denitrification depends on a combination of soil factors and is difficult to predict with certainty. The nitrate concentration, the degree of acidity or alkalinity (pH), the oxygen concentration and the amount of decomposable organic material all play a part. In general, strongly acid soils usually emit only nitrous oxide while non-acid soils in the temperate regions emit mainly nitrogen gas, except when there is a large concentration of nitrate in the soil, as happens when fertilizer is applied. The nitrous oxide to nitrogen ratio is relevant after nitrate has left the soil. One of the processes now used by some water suppliers in France and Germany to remove nitrate from water before it is distributed is biological denitrification in a digester (Chapter 9). Clearly the conditions in the digester have to be such that nitrogen gas rather than nitrous oxide is produced.

One final twist in the story should be mentioned, and that is that denitrification is not the only process in the soil that produces nitrous oxide. Some can be formed during *nitrification*. The amount of nitrous oxide produced in this way is thought to be small compared with that emitted during denitrification, but there is evidence that it becomes more important when nitrification occurs in soil that is wetter than normal.

Nitrogen fixation by micro-organisms

We saw earlier how some micro-organisms release nitrate from organic nitrogen while others immobilize it as organic nitrogen. We have also just seen that some microbes can release nitrogen gas into the atmosphere, so it may come as no surprise to learn that others trap nitrogen gas from the atmosphere. This trapping, or fixation as it is usually described, is very important. On a global scale about twice as much nitrogen is fixed by soil micro-organisms as is fixed chemically by the fertilizer industry (Table 4.2) and some nitrogen gas is oxidized when lightning passes through the air. In Britain and other industrialized countries, however, the fertilizer industry fixes the greater amount (Table 4.2).

Many organisms have the capability to draw nitrogen from the air and incorporate it directly into their tissues. The oldest class of such organisms is the blue–green algae. These also fix carbon from carbon dioxide through photosynthesis, and the ability to fix both carbon and nitrogen would seem

Table 4.2. Inputs of nitrogen to the land surface of the earth (after Jenkinson, 1990).

Inputs – world scale	
Biological fixation	140×10^6 t year^{-1}
Fertilizer	74×10^6 t year^{-1}
Lightning	8×10^6 t year^{-1}

Nitrogen balance for the UK in 1985	
Inputs	t N $\times$ 10^6
Imported human food	0.1
Imported animal food	0.1
Rainfall	0.3
Fertilizer	1.6
Biological fixation	0.3
Total	2.4
Outputs	
To sea	0.3
Exported food	0.1
Unaccounted for	2.1
Total	2.4

to give these algae a considerable evolutionary advantage. They are often among the first plants to colonize soil left bare by receding glaciers, but despite this they are not particularly common in most soils. Why do they not cover the earth? The answer lies in the nature of the process by which nitrogen is fixed.

Nitrogen is a relatively inert gas. (Because it makes up 78% of the atmosphere the world would be a very different place if it was strongly reactive.) Atmospheric nitrogen is dinitrogen, N_2, in which the two nitrogen atoms are held together by a triple bond, $N \equiv N$. Considerable energy is needed to break such a bond, more than is available in most biological processes. The organisms that fix nitrogen manufacture an enzyme called nitrogenase, which is one of the few able to act as a catalyst for the cleavage of the triple bond under the conditions that are found in the soil in the field. This enzyme will also catalyse the breaking of other triple bonds such as those in acetylene ($H-C \equiv C-H$) or cyanide ($-C \equiv N$). (This interesting observation led to the idea that this long-established enzyme might have evolved to detoxify cyanide in the early atmosphere of the planet, but this is now thought unlikely.)

The inert nature of nitrogen gas is not the only problem for the nitrogen-fixing organisms. Another is that the nitrogenase enzyme is readily deactivated by oxygen, so the organism has to maintain anaerobic conditions around the enzyme if it is to fix nitrogen from the air. This must be one of the reasons why the higher plants did not evolve the ability to fix nitrogen. What some plants did instead was to develop symbiotic relationships with bacteria that could fix nitrogen. These species include the leguminous plants such as clovers, lucerne (alfalfa), peas and beans, and also some trees such as alder, leucaena and acacia. The bacteria grow in nodules on the plant's roots, where they get a home and an environment in which they can control strictly the concentration of oxygen around the nitrogenase enzyme. They also receive a supply of carbon compounds from the plant from which they can obtain the energy that they need to split the triple bond of dinitrogen. In return they supply the plant with nitrogen that they have fixed from the atmosphere. These symbioses seem to be quite specifically between particular plant species and particular bacteria.

Much research is currently being devoted to the chemistry and biology of the nitrogen fixation process and the nature of the bacteria and their relationship with their plant hosts. There are obvious benefits from genetically manipulating an arable crop so that it can form nodules and enter into a symbiotic relationship with a nitrogen-fixing bacterium. Because of the specificity of such relationships we are still a long way from developing, for example, a wheat crop, able to form one. Also, a nitrogen-fixing wheat crop would not be the answer to all our problems. Nitrogen fixation costs the plant energy in the form of carbon compounds, so a wheat crop that relied on fixed nitrogen would not be capable of yields as large as those from a crop given fertilizer, and as we saw before, the fact that no fertilizer was used would not mean that no nitrate was washed out of the system. The microbes that decompose organic matter and those that convert ammonium to nitrate take no account of how the crop that has provided the residues received its nitrogen.

The leguminous crops make a substantial contribution to agriculture even though we do not yet have nitrogen-fixing cereal crops. In the temperate zones, clover plants are an essential component of grassland farming. The highly productive Canterbury Plains of the South Island of New Zealand depend almost entirely on nitrogen fixed by clovers. In hill country or other areas with difficult terrain clover is a far more practical option than applications of nitrogen fertilizer. Field beans and peas have an important role in European agriculture both as useful crops in their own right and also as 'break crops' between cereals that help to prevent the transfer of disease from one cereal to another and, as a bonus, leave some nitrogen for the next crop. In tropical areas and in the warmer parts of the temperate zones, soya and ground-nuts are both crops of huge economic and culinary importance. The nitrogen-fixing bacteria should certainly be counted among man's best friends; they play a vital role in feeding both ourselves and the animals on which we depend.

Chapter Five
Measuring Nitrate Losses from Soil

The nitrate debate needs facts, but the measurements of nitrate losses needed to provide them are difficult. Measuring nitrate concentrations in water is a matter of routine chemical analysis; the problem lies in collecting the water. The water we are interested in is that draining from the soil at the base of the root zone; that is the zone in which the crop's roots collect water and nitrate and below which we have to consider the nitrate as having 'escaped'. Different crops root to different depths, but for most of them the depth at which the measurement needs to be made is of the order of 1–2 m. We have somehow to collect the water that is moving through the soil at these depths. We also need to estimate how much water is passing through the soil. The reason for this is less obvious; it might reasonably be thought that all we need to do is to measure the *concentration* of nitrate in the water and check that it is at an acceptable level when it leaves the soil. We need to know too, though, the total *quantity* of nitrate leaving the soil in a given area and the quantity of water in which it is carried. The reason for this is that the water and nitrate from one area will eventually mingle with water and nitrate from other areas, and we can only work out the resulting concentration of nitrate if we know how much water and how much nitrate is coming from each area.

Methods for collecting water and the nitrate that it contains fall into four general categories, two of which have elements in common. We shall assess these methods by asking questions such as: How *easy* is the method to use and how *cheap* is it? How *reproducible* are the results? How much is the method affected by the fact that the structure of many soils results in the existence of (relatively) *mobile and immobile categories of water* (Chapter 3)? Is the *spatial variability* of the soil a problem? What exactly is measured by the method, and is this what we want to know – how relevant is the method? How easy is it to *interpret* the results in relation to the nitrate problem?

Extracting the soil solution by suction: porous ceramic cups

This approach has a fairly long history, going back to 1904 when two American scientists (Briggs and McCall, 1904) reported a device described as an 'artificial root'. This 'root' was a porous tube through which water could be drawn from the soil under suction. Its modern equivalent is usually described as a porous ceramic cup (Wagner, 1962). The cup is attached to a body made from plastic or metal tube and tubes are attached so that suction can be applied and the soil solution withdrawn into a collector (Fig. 5.1), once the assembly has been satisfactorily installed at an appropriate depth in the soil.

Porous cups are relatively cheap when the solute of interest is nitrate. A plastic tube can be used for the body and a sampling unit ready for insertion in the soil costs about £20 (US$40). Pesticides, however, tend to be adsorbed on the plastic tube, so a sampling unit for pesticides needs a

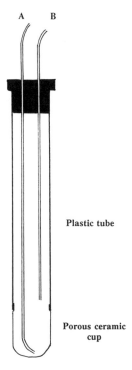

A   B

Plastic tube

Porous ceramic
cup

Fig. 5.1. A porous ceramic cup assembly. Air is withdrawn through tube B to create a partial vacuum that draws soil water into the porous ceramic cup. This water is collected by applying a vacuum to tube A. The pores of the ceramic cup are made too fine to admit soil particles.

stainless steel tube, which increases costs 10-fold. Porous cups are generally reckoned to be fairly easy to use, but whether they are easy to use *well* may be another question. Problems may arise during both installation and use.

The sampling units are installed in carefully prepared holes in the soil. Some users (e.g. Wagner, 1962) have relied on a 'snug fit' approach in which they make the hole marginally larger than the diameter of the unit. This is probably satisfactory when the units are installed horizontally from the side of a trench (Fig. 5.2), but the necessity of digging, and possibly shoring up, a trench detracts somewhat from the ease of use.

If, however, the units are installed vertically there is a risk that rain- or irrigation-water will be channelled down the remaining small gap around the sampling unit (Wood, 1973). Various approaches have been used to avoid the channelling problem. Some researchers sieve the soil and use it to back-fill the hole, while others back-fill with silica flour. Plugging the top of the hole with bentonite (a clay) also helps (E.I. Lord, personal communication); the bentonite expands greatly when wetted and seals the gap round the sampling unit. The units can be installed at 45° to the soil surface to lessen the risk of channelling, but they then need to be longer (Fig. 5.2). Another problem of installation is that of ensuring proper contact between the porous cup and the soil around it. Back-filling with soil helps, and silica flour may be even better. Another approach is to make a slurry of the soil taken from the hole and pour it back round the sampling unit.

Porous cups have not usually been found to be very satisfactory when the reproducibility of either the volumes of water extracted or their nitrate concentrations have been measured. Part of the problem comes from the intrinsic variability of the cups themselves, part from probable slight differences between the ways in which individual cups are installed, and the rest from the variability of the soil. Further problems arise when the porous cups are used; their pores become plugged with soil materials, and some

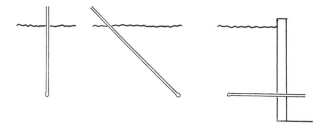

Fig. 5.2. Installing porous ceramic cup assemblies. Vertical installation is the most common but tends to channel water down around the assembly. This problem can be ameliorated by installing at 45° to the soil surface, necessitating a longer tube, or prevented by horizontal installation which means that an access pit is needed.

suffer more from plugging than others. Unfortunately, as we shall see shortly, the rates at which the cups fill can affect the nitrate concentration that they measure – and add to the problem of reproducibility. Two American researchers (Hansen and Harris, 1975) who made an extensive study on porous cups concluded that 'variabilities of ± 30% or more should be expected in field studies when sampling concentrations of 20 milligrams per litre of nitrate.'

Most of the original research on porous ceramic cups was made in the USA, but not all Americans were enthusiastic about them. One non-enthusiast (England, 1974) considered that 'in the present state of the art one cannot be sure from what macroscopic volume of soil the sample was extracted nor from which pores it was drained.' Whether or not the 'state of the art' has changed greatly in the past 16 years seems open to question. There are two questions here: what volume and which pores? The first was investigated in some calculations made by two German researchers working in the USA (Van der Ploeg and Beese, 1977). They concluded that porous cups could draw water from a sphere of about 0.6 m diameter; this might be seen as an advantage – the sample representing a fairly large volume – but it also has the problem that the sample is drawn from a wide variety of pores, not all of which may contribute to the flow. They also concluded that the suction applied through the porous cup could distort patterns of flow through the soil. That is, attempting to measure flows of water and nitrate in this way might alter what is being measured. Hansen and Harris (1975) showed that the faster water was drawn out of the soil through porous cups the smaller its nitrate concentration tended to be. The key point here is that the concentration *changed* as the water was drawn out faster; the fact that it became smaller probably just reflected the relative concentrations of nitrate in larger and smaller pores and in other circumstances it could probably have become larger.

This leads us to the second of the two questions: 'which pores'. In Chapter 3 we saw how structure of the soil influences the flow of water and nitrate through it. Water in very fine pores in soil aggregates moves very much more slowly than water in larger pores, to the extent that in soils with an appreciable degree of structure we can usefully think in terms of 'mobile' and 'immobile' water. We are interested in water that carries nitrate through and out of the soil, that is 'mobile water'. If porous cups draw water from a wide range of pores and not just those carrying this 'mobile' water how relevant are they to measurements intended to estimate the concentration of nitrate in water draining from the soil?

We saw in Chapter 2 that soils that are predominantly sandy have relatively little structure, while those that contain appreciable amounts of clay tend to be strongly structured. Silty soils usually lie between these extremes, as might be expected from the fact that the particle size range for silt lies between those of sand and clay. On this basis we could expect that porous ceramic cups would be most useful for sampling the water and nitrate likely to drain from sandy soils and least relevant in clay soil. This is just what was found by two American researchers (Barbee and Brown, 1986). They used porous cups to draw water out of soil, and also undermined the soil and inserted a horizontal collecting vessel to catch the water as it came through the soil. (More details of the second method are given later in the chapter) They compared the two approaches in three differing soils: a loamy sand, a moderately-structured silt loam and a strongly-structured clay soil. All these soils were in their natural condition in the field, except that they were kept bare of vegetation. The researchers applied a fairly concentrated solution containing chloride (500 mg l^{-1}) to the surface of each soil and followed its progress with the two collecting systems as natural rain washed it downwards. Samples taken weekly in the sandy loam soil showed no significant difference between the chloride concentrations measured in the water samples from the two systems [Fig. 5.3(a)]. Both systems also registered the peak concentration of chloride about 100 days after it was applied to the surface of the soil. This agreement did not, however, extend to the 'moderately-structured' silt loam [Fig. 5.3(b)]. Samples from both collecting systems suggested that the water was moving preferentially through larger pores in this soil, but while the porous cups recorded the peak concentration after only 8 days the horizontal collector recorded it at 42 days. The clay soil showed even larger differences [Fig. 5.3(c)]. Samples from the horizontal collector again suggested preferential flow, with the peak concentration coming at 28 days, 14 days earlier than in the silt loam and much, much earlier than in the sandy loam. The porous cups were not a success in this soil; they produced enough sample for analysis on only one occasion and seem to have been bypassed by the flow of water in the soil on other occasions. In general the horizontal collector yielded more sample more consistently.

Another group of American researchers (Shaffer *et al.*, 1979) found broadly similar results in clayey silt loam soil when they applied fairly concentrated solutions of nitrate to the soil surface. Concentrations of nitrate collected from the surface of a tunnel 1.2 m below the soil surface were up to 12 times greater than the concentrations in water extracted through a porous cup at 0.94 m depth (Fig. 5.4). In an additional twist to the experiment they spiked two of the nitrate solutions with cadmium. This came through rapidly to the roof of the tunnel but seemed to bypass the porous cup. Preferential flow through larger pores must again have been responsible, because cadmium is strongly adsorbed by the soil and does not leach in normal circumstances; such flow would have been encouraged by the relatively large (16.6 mm) aliquots of water applied. Such flow contributes increasingly to the overall flow as rain becomes more intense; so it is not surprising that porous cups also seemed to be bypassed during tropical storms in Costa Rica (Russell and Ewel, 1985).

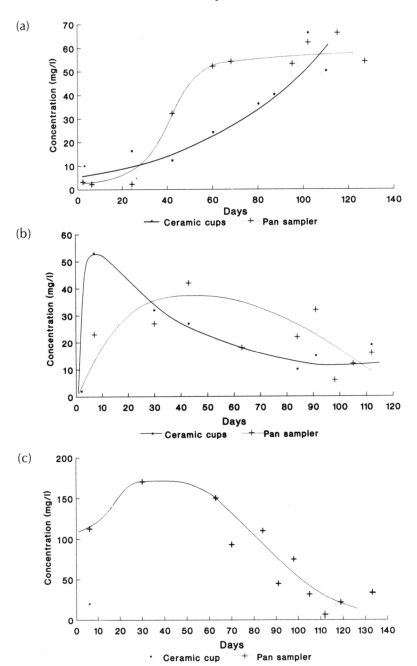

Fig. 5.3. Concentrations of chloride in soil water measured by porous ceramic cups (●) or a horizontal collector (+) in (a) sandy loam, (b) moderately-structured silt loam and (c) clay soils (Barbee and Brown, 1986).

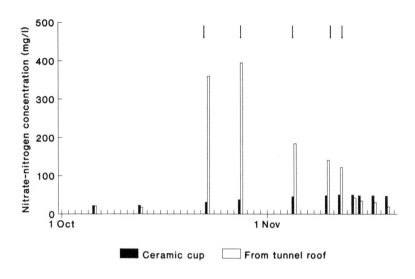

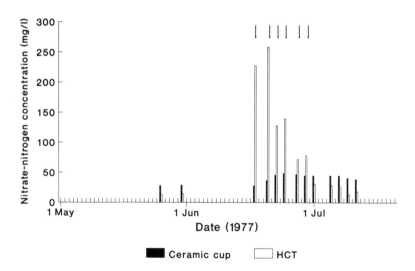

Fig. 5.4. Concentrations of nitrate in water passing through a clayey silt loam soil following applications of water to soil surface in 16.6 mm aliquots; water collected by porous ceramic cups, as drips at the tunnel roof or in a horizontal collector under tension (HCT) (Shaffer *et al.*, 1979).

It is tempting to simplify the results by saying that the porous cups were 'wrong' and the horizontal collection systems were 'right'. As we shall see, the latter are not without their problems, so we should not draw too facile a conclusion, but it is probably fair to conclude that porous cups are best restricted to sandy soils. They are relatively cheap and allow considerable flexibility, but much depends on the care with which they are installed and used. One practical point to remember is that water passes reasonably quickly through sandy soils – and also through large pores in some other soils. Measurements made with porous cups the day after rain or irrigation may be of rather less value than those made on the day!

Catching water as it passes out of the soil: lysimeters

The lysimeter like the porous ceramic cup, is quite simple in concept but not particularly easy to use. In essence, a volume of soil is held in position either by some form of retaining vessel or simply by the soil around it. Its upper surface is exposed to rain and evaporation in the usual way and may have plants growing on it. At the base of the soil there is some kind of system for collecting the water that has passed through the soil. The volume of water is measured, and sampled for chemical analysis if necessary. Some lysimeters can be weighed, usually to keep track of the amount of water held in the soil in them.

Lysimetry can trace its antecedents much further back than most branches of contemporary science – to 17th century Paris. A mathematician and meteorologist named De la Hire, who worked for Louis XIV, constructed three lead-walled lysimeters in 1688. His work was to remain unique for more than a century, until 1795–96 when two sets of lysimeters were established at about the same time but quite independently. One set was installed at Geneva by a Swiss named Maurice and the other in England by John Dalton. Dalton is best known as the 'Father of Atomic Theory', but he also kept a meteorological diary for 57 years and was interested in the origin of springs. These three sets of lysimeters, like all lysimeters constructed before 1870, were of the 'filled-in' type; that is to say, the soil was dug up and re-packed in the lysimeter. It later became apparent that, however carefully this was done, the flow of water in the repacked soil was not the same as in the soil in its natural state (Joffe, 1932). In some instances, for example, rain-water was simply not able to penetrate the re-packed soil and remained as a pond on the surface.

The first lysimeters to contain undisturbed soil were the 'Drain Gauges' constructed in 1870 by J.B. Lawes and J.H. Gilbert, the founding fathers of Rothamsted Experimental Station. These consisted of natural blocks of soil, with a surface area of 4 m^2, that were isolated by brick walls and undermined so that vessels to collect the drainage could be installed (Fig. 5.5). The significance of *undisturbed* soil was soon emphasized by the fact that Lawes and Gilbert were able to deduce from the results the existence of the mobile and immobile categories of water in the soil discussed above and recognize their significance in relation to losses of nitrate from the soil (Lawes *et al.*, 1882). They made it clear that the structure of the undisturbed soil was important, and these results should arguably have ensured that filled-in lysimeters were no longer to be used as a measure of water and nitrate movement through natural soil, but this was not to be so. The rest of this chapter, however, is concerned solely with lysimeters with undisturbed soil.

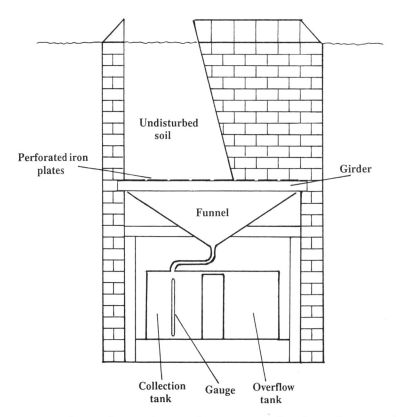

Fig. 5.5. A Rothamsted 'Drain Gauge'. Three gauges each with a surface area of 4 m², were constructed in 1870 to measure the quantities of water draining through 20, 40 and 60 inches of soil (0.5, 1.0 and 1.5 m) by Lawes and Gilbert in 1870. These consisted of natural blocks of undisturbed soil that were isolated by brick walls and undermined so that collectors for the drainage could be installed. The soil in each gauge carried no crop and received only minimum cultivation to kill weeds.

There are two ways of constructing a lysimeter with undisturbed soil: one is to leave the soil where it is, dig a trench beside it and insert some form of collecting vessel to catch the water passing down through the soil; the other is to remove the soil literally *en bloc*, with the minimum of disturbance, and attach some form of collecting system to the base of the resulting column of soil. The first approach leads to an *Ebermayer lysimeter*, so named in honour of a Bavarian meteorologist who, towards the end of the 19th century, planned to undermine and collect drainage from a quarter of a hectare (about 0.6 acre) of forest, complete with trees, at a depth of 1.5 m. This ambitious project failed to come to fruition because of the technical and financial difficulties, but the name lives on. The second approach leads to a 'monolith' or 'block' lysimeter. The key scientific difference between the two categories is that of lateral constraint. An Ebermayer lysimeter has no side walls, except possibly where the trench is, whereas a monolith lysimeter does have some form of side wall. Hard and fast distinctions are not always possible; the Drain Gauges, to which reference was made above, were built with side walls but the collection of the drainage is broadly similar to that in an Ebermayer lysimeter. From a practical point of view, however, it is convenient to divide the rest of this section along these lines.

Ebermayer lysimeters

Making a lysimeter of this kind involves digging a trench and inserting a collecting vessel horizontally (Fig. 5.6). The cost and effort involved depend on the depth at which the collector is installed. The cheapness of the equipment seems to depend on the ingenuity of the experimenters. Barbee and Brown (1986) made their collecting vessels by drilling holes to allow the entry of water through the top of a large hollow glass brick and inserting a nylon tube through which the water sample could be drawn out. Others have used stainless steel pans as collectors. To judge by published reports these lysimeters are reasonably easy to use, but, as with the porous cups, using them well needs attention to detail. Several authors emphasized two key points: making sure that the hole made in the wall of the trench for the collecting vessel is exactly the right size for it; and ensuring that the vessel makes proper contact with the soil above it. Both these points have parallels in the use of porous cups. There is a further important question, whether or not suction needs to be applied to the collectors, that is discussed later in relation to both categories of lysimeter.

We have not found any reports in the scientific literature in which these lysimeters had been tested for their reproducibility. The main factor would probably be the variability of the soil. They have the advantage over porous cups in that they are much less likely to be bypassed by water moving preferentially through large pores. When we turn to questions such as

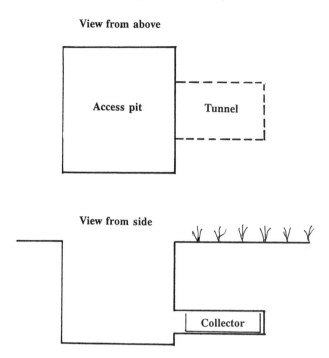

Fig. 5.6. The essentials of an Ebermayer lysimeter. A horizontal tunnel is made from an access pit and a collector inserted. The collector may be a porous plate to which tension can be applied.

relevance and interpretability we come up against the point that Ebermayer lysimeters have no lateral constraint – no side walls. This is both a strength and a weakness; a strength because any lateral flow of water that occurs in the natural soil is retained, and a weakness because there is no way of knowing exactly from what area the water and nitrate collected have come from. The latter would be a problem if differing treatments were to be applied to adjacent lysimeters.

Monolith lysimeters

Collecting a soil monolith involves driving a metal or fibreglass pipe into the soil, partially excavating around it, cutting through the soil at the base of the pipe and lifting out the pipe and the soil it contains. The monolith becomes a lysimeter when a water collecting system is attached to the base of the pipe (Fig. 5.7) and the whole system is either reinstalled *in situ* or removed for experiments elsewhere, possibly to become part of a 'bank' of lysimeters. These systems seem to occupy an intermediate position in terms of cheapness and ease of use. Our colleague Keith Goulding estimates the

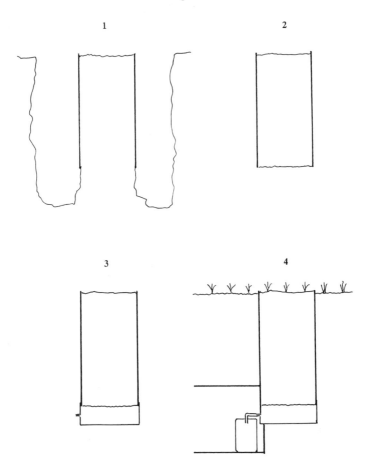

Fig. 5.7. Construction of a monolith lysimeter. (1.) A metal or fibreglass pipe is driven into the soil, and the soil around it is excavated as it moves down. A metal cutting ring (not shown) is usually attached to the lower edge of the pipe. (2.) The soil at the base of the pipe is cut through, as cleanly as possible, and the resulting soil monolith is lifted clear. (3.) A collector for water is attached. (4.) The resulting lysimeter is installed and attached to a vessel for the drainage. A system for applying tension at the base of the lysimeter may also be installed.

current cost (labour and materials) of a single monolith lysimeter to be about £2000. One measure of 'ease' is that a cylindrical monolith 0.8 m in diameter and 1.35 m deep weighs about 1.4 t! Despite this a team of researchers from the former Letcombe Laboratory acquired 64 monoliths for one project and reported that their collection was 'straightforward' (Belford, 1979). This should perhaps be seen as a measure of their skill and determination rather than the underlying ease of the technique. We need to ask again whether they are easy to use well. Possible problems identified by the Letcombe team included compression of the soil, vertically and also laterally next to the pipe that forms the casing, and unstable soils whose structure collapses during the collection of the monolith. They also emphasized the importance of getting a clean break at the base of the soil.

The reproducibility of the results from monolith lysimeters, like that from Ebermayer lysimeters, will often depend more on the variability of the soil than that of the technique. Preferential flow through larger pores in the soil is again no problem; water cannot bypass the collection system at the base of the monolith. There can, however, be a serious problem with a clay soil that shrinks away from the wall of the lysimeter during dry weather; water may then bypass the entire block of soil. Turning to the questions of relevance and interpretability raises the question of lateral constraint – once again both a strength and a weakness. There is little doubt as to the area of soil from which the nitrate and water have been collected, but the lateral flow is prevented. This is not likely to be a problem in sandy soils or in other soils through which water can pass reasonably freely.

Ensuring the soil condition at the lysimeter base is as natural as possible

The surface tension of water has a strong effect on its behaviour anywhere that water has an interface with air. It is responsible, for example, for the hemispherical shape of a drop of water on glass, and as we saw in Chapter 3 it causes pressure differences that drive water into fine pores in the soil. It is also a factor that we need to consider when using lysimeters.

The pores in the soil that conduct water downwards are usually reasonably continuous. This means that here is a 'hanging column' of water that extends to depth without a horizontal interface between water and air. This 'hanging column' is broken when a lysimeter is constructed. The surface tension at the resulting air–water interface holds back drainage of water from the soil, so that water drains from the soil above the interface only when the soil becomes saturated (Richards *et al.*, 1939). This problem can be avoided by applying suction to the base of the soil, and it has been shown that the suction controls both the rate at which water drains from the soil and the amount of water held in the soil at the base (Coleman, 1946). Pores of different sizes drain at different suctions (Chapter 3). This means that the suction determines which particular suite of pores is releasing water and this could influence both the quantity and the concentration of nitrate in water leaving the soil. This is not just hypothesis. Three American researchers (Haines *et al.*, 1982) compared two Ebermayer-type lysimeters, both at the same depth in the same soil. One had the collector subjected to a 1 m hanging column of water; the other had no suction applied to it. The average flow of water was twice as large when suction was applied as when it was not, and, more critically, the average nitrate concentration was three times smaller (Table 5.1). The authors concluded that applying suction made the sampling of the smaller pores more efficient and that of the large pores less so. The reason that the suction made the nitrate concentration smaller is probably that the nitrate in this particular forest soil came from decomposing organic debris on the surface, rather than from the main body of the soil; this nitrate was brought down when there was enough rain to cause flow in the larger pores. The smaller pores which drained under suction probably contained less nitrate. In a different soil, one under arable cultivation perhaps, the smaller pores could contain a larger concentration than the larger ones, so that the ratio of nitrate concentrations would be reversed. The key point is that there can be a difference.

It seems clear that it may be important to apply suction to lysimeters, but how should we apply it and how much should be applied? The most common approach in Ebermayer lysimeters seems to be to place a porous ceramic plate firmly in contact with the base of the soil and apply suction to this, but there are other options. Shaffer *et al.* (1979) applied the tension through a wad of filter paper held in a special funnel. The options suggested for monolith lysimeters by researchers at Letcombe (Belford, 1979) were inserting porous candles at the base of the soil or keeping the base in contact with a matrix of silica sand; both allow suction to be applied. It is not really possible to generalize about how much suction should be applied. This depends on the length of the hanging column that the soil would experience in its natural state.

One point that may have occurred to you is this. The critique that was made of porous ceramic cups earlier in the chapter was mainly made in terms of comparisons with lysimeters to which no suction was applied. Could the absence of suction have invalidated these comparisons? Two

Table 5.1. Water flows and nitrate concentrations in lysimeters with and without tension (Haines *et al.*, 1982).

	Tension	No tension
Mean water flow per sampling (mm)	67	32
Mean concentration nitrate–nitrogen (mg l^{-1})	0.9×10^{-2}	3×10^{-2}

Note: These very small nitrate concentrations were in an uncultivated, unfertilized forest soil.

considerations suggest that this was not so. One is that the differences in heavier soils between the porous cups and the lysimeters without suction were generally larger than those between the lysimeters with the without suction. The other one is that one of the research groups in America (Shaffer *et al.*, 1979) had the foresight to use lysimeters with and without suction in the comparisons with the porous cups (Fig. 5.4). The nitrate concentrations in the water sampled by the porous cups differed substantially from those in water collected in the lysimeters whether or not suction was applied to the lysimeters.

Taking advantage of impermeable subsoils: collecting water from field drainage systems

Heavy clay soils impede the downward flow of water, except when they dry out and form cracks. This means that some form of artificial drainage is often put in to make them useful for farming. A system of 'mole' and tile drains is often used. The 'mole' is a piece of metal, shaped as you would expect from the name, that is drawn through the soil on the end of a metal bar. The mole creates a small drainage channel, while the bar cuts a slit in the soil and leaves a 'herring-bone' network of cracks on either side of it that helps to collect surplus water. Running at right angles to and slightly beneath the mole drains are tile drain pipes made from porous material that are placed in a prepared channel and covered with gravel before the soil is replaced above them (Fig. 5.8). This kind of drainage system is usually put in only when the soil beneath it is almost impenetrable to water so that drains collect a large proportion of the water passing through the upper part of the soil. There are three main ways in which water and solute move in these soils: surface runoff, that is, water movement across the surface of the soil; interflow, which is lateral flow on the top of the impermeable subsoil beneath the plough-depth; and flow into and through drainage systems such as that described above. Downward seepage below the drains usually contributes little to water movement.

Surface runoff, interflow and drainflow can all be collected, measured and analysed, which makes such systems a useful tool for assessing the movement of nitrate in the soil – a fact that did not escape Lawes and Gilbert at Rothamsted in the last century; they put drains under some of the plots of the famous Broadbalk experiment (see Chapter 7). The Broadbalk plots, however, did not have any facilities for collecting surface runoff or interflow, and there were no physical barriers to prevent lateral flow between plots. This may not have been much of a problem because water usually moves downwards quite freely in the soils at Rothamsted. A more recent experiment, the Brimstone experiment, named after the farm

Soil

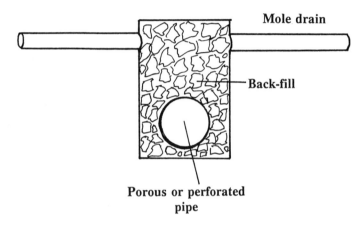

Mole drain

Back-fill

**Porous or perforated
pipe**

Fig. 5.8. Mole and pipe drains. A mole drain is made by drawing a mole-shaped piece of metal through the soil. It interconnects at right angles with a porous pipe drain which lies just below it in a trench back-filled with large gravel.

in which it was established, was designed by the Letcombe research team and the Field Drainage Experimental Unit to have these features (Cannell *et al.*, 1984). This experiment involved substantial inputs of time and effort and money and it has few counterparts elsewhere. Two known to us have been set up at Uppsala in Sweden (Brink and Linden, 1980) and at Jokioinen in Finland (Jaakkola, 1984).

These experiments have plots that are large by comparison with most lysimeters. This means that the site has to be selected carefully to avoid major differences in the soil between plots, but preferential flow is no problem at this scale and much of the soil variability is also within the scale of the plot. Another big advantage is the relevance of such experiments to normal agriculture. The soil is largely as it would be in an ordinary field. Lateral flow is no problem because it is collected and analysed, and any effects of air–water interfaces that occur in these drainage systems are similar to those that occur in most under-drained fields. Water drains from these plots very much as it would elsewhere. Experiments of this kind are also readily interpretable. The area of soil from which the water and nitrate come is clearly delineated by plastic barriers and the concentrations and losses of nitrate can be determined for all three types of flow: surface runoff, interflow and flow from the drains. These systems therefore give a very clear idea of where the water and nitrate come from and where they go and they also allow the greatest degree of realism with respect to farming practice.

Sampling the soil rather than the water

One other approach needs to be mentioned in passing. Sampling the soil is a cheap option and may be an easy one, provided there are not too many stones in the soil. It obviously does not tell you directly how much nitrate is leaving the soil or at what concentration, but it does give a measure of the amount of nitrate being washed out. Successive samplings give information on changes in the amount of nitrate in the soil but these changes do not necessarily arise from downward movement. Crops remove nitrate and microbes in the soil may transform it – or produce it.

One possibility is to use a computer model (Chapter 6) that simulates the changes brought about by these various processes, including that of washing out (e.g. Addiscott and Whitmore, 1987). If the model is successful in simulating the changes in the amount of nitrate in the soil, some credence can be given to the estimates that it gives of the amounts and concentrations of nitrate leaving the soil (e.g. Powlson *et al.*, 1989) Caution is necessary, however; if, for example, the model has compensating errors in two of its routines, it could possibly give a convincing simulation of the amounts of nitrate in the soil while giving a completely misleading estimate of the quantities of nitrate being lost at the base of the soil. This approach is probably wise only if the user is familiar with the model and the model has been tested previously on the same or a similar soil.

Is there a 'best buy'?

Anyone reading this chapter as a kind of consumer report on the various methods has to face the fact that there is no real 'best buy'. Much depends on the type of soil studied and the resources available to the investigator.

Those working with limited resources on fairly uniform unstructured soils may find that placing porous ceramic cups near the base of the rooting zone is the best option. There will be some uncertainty as to how well the water sampled by the porous cups represents the water that is flowing through and out of the soil but the concentrations of nitrate found in the samples will probably not differ greatly from those in the mobile water. Two much larger problems arise if the experimenter also wishes to estimate the flux of nitrate from the soil. The first is that he or she will have to multiply the concentration of nitrate, which is known, by the flux of water, which is not known. Estimating fluxes of water through the soil is notoriously difficult. The best hope is probably to subtract an estimate of the amount of water evaporating from the soil and crop from a measurement of the amount of rain or irrigation. The other problem is that if the concentration of nitrate measured varies from point to point, as it will do, and the flux of water also varies, there is some question as to whether multiplying the mean concentration of nitrate by the mean flux of water gives the true mean of the flux of nitrate. Some account may need to be taken of the extent of the variability. Anyone contemplating a calculation of this kind might do well to consult a statistician first.

Lysimeters seem on balance to offer a more reliable guide than porous ceramic cups to the concentrations of nitrate in water leaving the soil, but thought must be given as to whether suction should be applied to the soil. If it is, it should ideally be applied to the base of the whole lysimeter and correspond with the suction applied by the 'hanging column' of water that the soil would experience in its natural environment. The experimenter also needs to choose between an Ebermayer lysimeter and a monolith lysimeter with side walls. The decision depends on the degree of priority attached to: (a) retaining a natural flow of water, including any lateral flow; and (b) being certain where the water and nitrate collected came from. The second point becomes important when the flux of nitrate needs to be known; the area of soil from which the nitrate has come is defined by the walls of the lysimeter.

Finally, if you have an impermeable clay subsoil and a quarter of a million pounds to spend, your 'best buy' is unquestionably a large-scale drainage collection system like that at Brimstone farm.

Chapter Six
Chasing Nitrate with a Computer: Modelling

We mentioned computer models in earlier chapters and in this one we describe various kinds of computer model and how they are used – and also abused.

We need first to say exactly what we mean by a computer model. Computer models are used to describe processes ranging from the flow of traffic in cities, to the generation of revenue from taxation to the degradation of material by microbes. In each case the model is a *representation* of the reality, just as a model ship is a representation of a real ship.

In a scientific context a model can also be seen as an *extended hypothesis*. There is a great deal of argument about how science, and scientists, actually function, but one of the more useful descriptions (or models) is the hypothetico-deductive principle of Karl Popper (Popper, 1959; Medawar, 1967). Popper argued that science advances through scientists forming hypotheses and testing them against reality, in the form of a critical experiment that can refute the hypothesis if it is not correct. If, as is often the case, the hypothesis is not correct, it can be revised and re-tested until it matches reality sufficiently closely. The match with reality must be assessed in some objective way, which often means using statistics. The statistician usually tests the 'null hypothesis', which involves calculating from the experimental data the *probability* that the hypothesis advanced by the scientist is incorrect; the smaller this probability, the more satisfactory the hypothesis.

A hypothesis can be in words. One famous example suggested that the earth went round the sun, rather than the sun round the earth as had been assumed previously. This was tested against astronomical observations and found correct. The most elegant hypotheses are usually simple; Einstein's famous equation relating energy (E) and mass (m) through the speed of light (c), $E = mc^2$, is an obvious example. In soils and many other natural systems, however, the reality is too complex to be expressed with such elegance. Nitrate is a prime example. It may descend in rain or be washed out by rain; soil micro-organisms produce it, immobilize it and destroy it; plants take it up and the farmer applies it in fertilizer. As all these processes can happen practically at once there is no single hypothesis, and the complexity that results from any attempt at a simultaneous description of the processes requires the intervention of a computer. Hence too the term 'extended hypothesis'. As we shall see, these processes can be described with various levels of exactitude, all of them imperfect. The extended hypothesis, like all hypotheses, has to be tested against reality. In the current parlance, models must be *validated.* One of the uses to which models are frequently put is the prediction of nitrate leaching and other processes in circumstances in which they cannot be measured. It is obviously important that models should be properly validated before this is done. Indeed, the use of a model to make predictions that are to be acted upon in a context for which it has not been validated in an appropriate way constitutes an abuse. This is particularly serious when government policy is involved, and we felt called upon to comment on one such case recently (Addiscott and Powlson, 1989). Other uses to which models can be put include the exposure of areas of ignorance and the extraction of extra information from experimental data. An example of the latter is given on p. 100.

Some people are not enthusiastic about models and modellers, usually because they feel that modellers fall short of scientific rigour. One such is the distinguished Australian soil physicist John Philip, who expressed some fairly pungent views on modelling in a very entertaining way in a recent article (Philip, 1991). We agree, as we hope we have made clear, that models must be used in a proper way.

As you will have gathered from the other chapters the leakage of nitrate is governed by three main processes.

1. Water flowing through the soil, which tends to wash the nitrate out.
2. The activities of the microbes in the soil, which produce nitrate as they break down nitrogen-containing organic material in the soil. They can also remove nitrate from the system by locking it up in soil organic material or by converting it to gaseous nitrogen or nitrous oxide.
3. Crops growing on the soil, which take up nitrate and remove it from the risk of leaching.

These three processes differ considerably in nature and need different modelling approaches, but they cannot be treated in isolation. Nitrate produced by the microbes will be washed out by rain-water unless it is taken up by the crop; all three processes therefore need to be modelled simultaneously as far as possible. They also interact in other ways. A shortage of water in the soil, for example, will hold back both the production of nitrate by the microbes and its uptake by the crop, while a shortage of nitrate in the soil obviously restricts uptake by the crop.

We shall consider models for each of the three processes separately and discuss later how they can be combined. One question that comes up with models for all three processes is how great a degree of complexity is productive. We shall describe first models for the physical side of nitrate behaviour, the washing out of nitrate by water passing through soils.

Modelling the wash-out: physical leaching

Many models of leaching have been developed over the years and we cannot describe them all, but shall point out the key distinctions between the different kinds of model and show how these relate to the uses to which the models are put. These distinctions were discussed in greater detail by Addiscott and Wagenet (1985).

The first is distinction between *functional* and *mechanistic* models. Functional models aim to give a reasonably good general description of the leaching process and its results without going into great detail. Mechanistic models, on the other hand, try to incorporate the best possible description of the process in terms of classical physical theory. This means that they are more complex and employ a higher level of mathematics, usually differential equations.

The next distinction is between *capacity* and *rate* models. The capacity of the soil to hold water is important for the simple reason that the more water the soil can hold, the more has to fall in rain or be applied in irrigation to displace what is there already (Chapter 3). Rate in leaching models means the rate at which the soil transmits water, that is, its conductivity to water. This influences the quantity of water that is transmitted and thence the quantity of nitrate that is washed out.

We need here to introduce the word 'parameter', which features extensively when we talk about models. The Oxford dictionary says: '**parȧ´meter** *n.* (Math.) quantity constant in case considered, but varying in different cases'. A parameter of a leaching model is a soil property which: (a) influences leaching; and (b) is more or less constant for a particular soil, but varies between soils. The distinction between rate and capacity models is better expressed as that between models that depend mainly on *capacity parameters* and those that depend on *rate parameters*.

The next distinction is between *deterministic* and *stochastic* models, and is definitely the most difficult. A deterministic model presumes that a given set of events leads to a uniquely definable outcome, while a stochastic model presumes the outcome to be uncertain and is structured to account for this uncertainty.

The final distinction is that of *use* or *purpose*. Some models are primarily tools for *research*, developed to help test ideas and expose areas of ignorance, while others are *management* aids for the farmer and his advisers. These distinctions overlap to a substantial extent. In general, functional models use rate parameters and fall into the management rather than the research category, while mechanistic models use rate parameters mainly in a research context. The parallels, however, do not extend to the distinction between deterministic and stochastic models.

Some individual leaching models: How they work and how they perform

The simplest model assumes that the incoming water or solution pushes out the solution already in the soil pores in a completely uniform way and without mixing with it. This is 'piston flow', so called by analogy with a piston pushing out or displacing, a fluid from a cylinder. It simply says that the depth, z_p, to which the front of the displacing solution penetrates depends on the quantity, Q, of the water or solution that is doing the displacing and the volumetric moisture content, θ, which is the fraction of the soil's volume that can hold water.

$$z_p = \frac{Q}{\theta} \qquad\qquad (6.1)$$

Obviously z_p and Q have to be in the same units. The more water the soil can hold the smaller is the downward movement caused by a given amount of rainfall or irrigation. Because this model assumes that the displacement is uniform it is not satisfactory in soils in which some of the soil water is more mobile than the rest, and it works best in sandy soils. Two other limitations need mention. One is that although the model gives the position of the front of the displacing solution, it gives no idea of how much this front is spread out. The other is that the model can only deal with a single application of nitrate. This may be appropriate for an application of fertilizer, but we are often concerned with nitrate fed more or less continuously into the system by the microbes in the soil. You will probably have deduced that this is a functional model, whose parameter, θ, is a capacity parameter. The quantity, Q, of water does not count as a parameter; θ is assumed constant for a particular soil but not for all soils, while Q is not even a constant – it depends on the whims of the weather and the experimenter. This model is most likely to be used in a deterministic way and for management rather than research.

A useful leaching equation was developed at the former National Vegetable Research Station, now Horticultural Research International (Burns, 1975). This gives the fraction, f, of nitrate or a similar solute that is washed below a depth h cm by the percolation of P cm of rain:

$$f = [\frac{P}{P + \theta}]^{h} \tag{6.2}$$

This equation applies to an application of nitrate to the soil surface. If the nitrate is initially distributed through the soil to the depth h, the h in the equation is replaced by $h/2$. (You can see this as nitrate starting on average half-way down, but Burns presented some more mathematically-based reasoning.) This equation seemed to apply reasonably well to data from a number of locations (Fig. 6.1) and it has the advantage that it is simple enough to be used on pocket calculators. It was also incorporated into a simple leaching calculator for growers. By using a succession of values for h you can get an estimate of the way the nitrate has become spread downwards through the soil, but the equation again cannot cope with nitrate fed into the system by microbes. You should by now have listed against this equation the words, functional, capacity, deterministic and management.

A useful functional model was developed again by Burns (1974). It can be pictured (Fig. 6.2) as a vertical series of stirred containers each of which overflows into the one below. Water added to the top one is assumed to mix instantaneously with water and nitrate already there before an equivalent volume overflows. These containers represent layers in the soil and the volume held in each of them corresponds to the volumetric

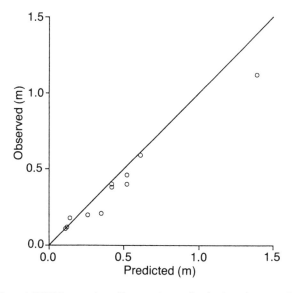

Fig. 6.1. Burns' (1975) equation. Comparison of calculated mean displacements with those measured.

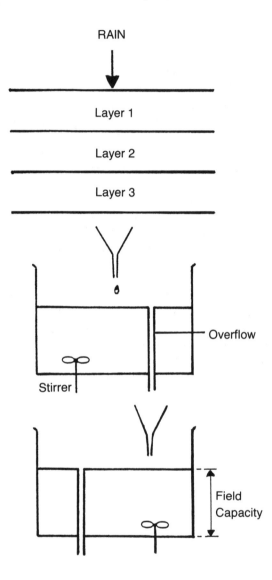

RAIN

Layer 1

Layer 2

Layer 3

Overflow

Stirrer

Field Capacity

Fig. 6.2. Pictorial representation of Burns' (1974) model. Water added to the stirred container at the top mixes instantaneously with water and nitrate already there before an equivalent volume overflows and the process is repeated in the subsequent containers. The containers represent layers in the soil. (Diagram courtesy of Dr Ian Burns.)

moisture content, θ, of the soil at 'field capacity'. Field capacity is the condition of the soil when any surplus water has drained but no water has yet been lost by evaporation. This is a rather ephemeral state that is not readily measured in clay soils because water drains so slowly from them. (We should now add that in the Burns' equation cited above θ should strictly be θ_{fc}, that at field capacity.) This model also allows for the drawing upwards of water and solute through the soil that occurs when water is evaporated from the surface of the soil. All these events are described in a program written in the computer language Fortran. This model was shown to be able to simulate satisfactorily the movement of nitrate and chloride under natural rain and evaporation in a soil in a field. The simulation for chloride (Fig. 6.3) was the better, probably because the microbes in the soil removed some of the nitrate. This model is able to show how nitrate or chloride is distributed through the soil and it is capable of describing the leaching of nitrate fed into the soil by the microbes. Neither Burns' equation nor his model takes account of mobile and immobile water in the soil and are therefore likely to perform best on sandy or other relatively unstructured soils. The model falls into the same categories as the equation, that is, functional, capacity, deterministic and management. Its main parameters are θ_{fc} and θ_{el}, which is the value of θ at the evaporative limit, the point at which no more water can be removed by evaporation.

A functional model for soils with mobile and immobile water

Chapters 3 and 5 showed that many soils have enough structure to make it necessary to consider both mobile and immobile categories of water. Models developed by Addiscott (1977) and Addiscott and Whitmore (1991) aim to represent this phenomenon. They fall into the same category as the Burns' model and, like his model, they divide the soil into horizontal layers. The key difference is that the water in each layer is then divided into mobile and immobile categories (Fig. 6.4), and rain or irrigation causes flow through the soil in the mobile water. After the flow, solute moves between the mobile and immobile water such that it tends to equalize the concentrations between the two categories.

This model has existed in a number of versions. In the first [Addiscott, 1977; Fig. 6.4(a)] the amounts of water, w_m and w_r, in the mobile and immobile (or retained) water, respectively, were fixed for any particular soil and were therefore parameters. In the most recent, acronym SLIM [Addiscott and Whitmore, 1991; Fig. 6.4(b)], w_r remains a fixed parameter for any particular soil but w_m is allowed to vary and a new parameter, α, is introduced. This α is the proportion of the mobile water in any layer that moves to the next layer in any one day. Thus α is a measure of the ease with which water flows through the soil and is effectively a rate

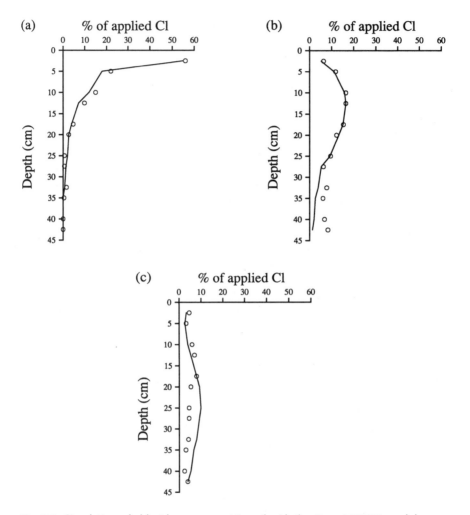

Fig. 6.3. Simulation of chloride movement in soil with the Burns' (1974) model. (o) Measurement; (——) simulation. Chloride applied 14 May 1970. Profiles at (a) 13 July 1970, (b) 3 August 1970, (c) 20 October 1970. Redrawn from Burns (1974).

parameter; it was shown in one study to be related to a measurement of the infiltration of water. This model is therefore possibly the only leaching model with both a capacity and a rate parameter. Both versions of the model have proved successful in simulating measurements of nitrate and chloride leaching in the field. The first was used to simulate the concentrations of chloride measured in water draining from the soil of the Drain Gauges at Rothamsted (see Chapter 5). The concentrations showed a very distinctive pattern that the model simulated to a large extent (Fig. 6.5). The

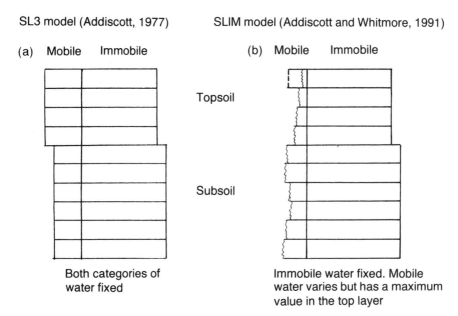

SL3 model (Addiscott, 1977)　　　SLIM model (Addiscott and Whitmore, 1991)

(a) Mobile   Immobile　　　　　(b) Mobile   Immobile

Topsoil

Subsoil

Both categories of
water fixed

Immobile water fixed. Mobile
water varies but has a maximum
value in the top layer

Fig. 6.4. The layer model with mobile and immobile categories of water. (a) The earlier (SL3) model (Addiscott, 1977); (b) the SLIM model (Addiscott and Whitmore, 1991).

pattern reflects the influence of the mobile and immobile water. This family of models was developed with the management of solutes in view, which ties in with their place in the 'functional' category of models. The second model has been used with its parameters presented in a way that takes account of their natural variability. This exercise showed that taking the variability into account improved the simulation of the data, but the improvement was not so large as to suggest that ignoring the variability would cause serious errors.

The convection–dispersion model is the classic *mechanistic* approach to modelling the transport of nitrate or other solutes through the soil by water, and its main parameters fall into the 'rate' category. Convection or mass-flow, implies that the solute is carried in the flow of water. Dispersion means that the solute becomes dispersed, or spread out, within the flow. There are two reasons for dispersion happening (see also Chapter 3). One is that water flows more rapidly in some pores than others, so that the solute is carried forward faster in some places than others. The other lies in diffusion, the natural tendency of a solute to spread itself uniformly throughout the whole volume of water in which it finds itself. Diffusion occurs whether or not the water is flowing.

These concepts are more sophisticated than those we discussed in the

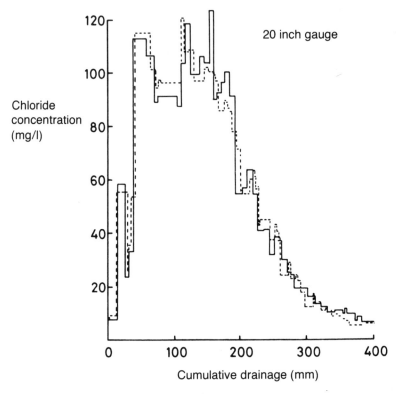

Fig. 6.5. Simulation of the leaching of chloride from the 20 inch (0.5 m) Drain Gauge at Rothamsted under natural winter rain. The complex pattern results from the interaction of the pattern of rainfall with the mobile and immobile categories of water that result from the structure of the soil. — measured, ---- simulated.

other models, but how easy are they to apply? Much depends on the nature of the flow. 'Steady-state' flow is the term used to describe what happens when an already wet soil experiences a continuous and unvarying flow of water, usually downwards. Neither the flow nor the amount of water in the soil changes with time. The mathematics considers the flux of solute and the flux of water, that is the quantity of solute and the volume of water crossing unit area in unit time (Wagenet, 1983). A flux is often represented by the letter J, so we have for solute and water J_s and J_w. If the flux of solute, J_s, is a 'steady-state' flux like that of water we have the fairly simple equation.

$$J_s = [\theta D_m(J_w) + D_p(\theta)] \frac{dc}{dz} + J_w c \qquad (6.3)$$

in which: θ is the volumetric water content as before, c is the concentration of solute and z is the depth from the surface of the soil. D_m is described as the mechanical dispersion coefficient, and reflects the effects of differing pore sizes as described above; these effects depend on how large the flow is, so D_m has to be presented as $D_m(J_w)$, the value for the given water flux. D_p is the diffusion coefficient, and since diffusion depends on how moist the soil is, it is presented as $D_p(\theta)$.

If the amount of solute changes at any depth, z, in the soil the flux of solute, J_s, is no longer a steady-state flux and we need a new formulation that defines the rate of change with time, t, of the amount, s, of the solute at depth z:

$$\left(\frac{\partial s}{\partial t}\right)_z = -\frac{\partial}{\partial z}\left(-[\theta D_m(J_w) + D_p(\theta)]\frac{dc}{dz} + J_w c\right) \tag{6.4}$$

This equation, like the previous one, is reasonably manageable because it is amenable to an analytical solution; that is, it can be manipulated into a form that gives direct answers when the appropriate parameter values are supplied. It is therefore a useful equation – provided there is steady-state water flow in the soil. This state of affairs usually happens, however, only in experiments specially designed to test equations for steady-state water flow! With nitrate leaching we are usually concerned with non-steady-state flow, that is, flow in which both J_w and θ change with time. This introduces a very important complication – the need to calculate J_w and θ at each time interval, using Equation (6.5) in which K is the hydraulic conductivity and H the hydraulic potential (Wagenet, 1983):

$$\frac{\partial \theta}{\partial t} = \frac{\partial}{\partial z}\left[K(\theta)\frac{\partial H}{\partial z}\right] \tag{6.5}$$

For the flow of water in soils K and H are not independent. The conductivity K depends on θ [hence $K(\theta)$] and θ depends on H. A soil physicist will say that the hydraulic properties of the soil have to be defined in terms of the $K - \theta - H$ relationship. Using this approach to obtain values for J_w and θ at various depths in the soil is not simple and the resulting value for J_w still has to be used to compute the flux of solute J_s from Equation (6.4). It is no longer possible to find an analytical solution for the combined equations and the modeller has to resort to forms of computation known as finite difference and finite element methods. The fact that the finite element method was developed to solve complex differential equations for the American aerospace programme gives some idea of the level of complexity involved.

Even if the modeller manages to get all these equations working together satisfactorily, his problems are not at an end; he has yet to deal with the problem of spatial variability. The parameters that link K, θ and H are not unique. They vary appreciably from point to point in a field and really need to be presented to the model in a way that takes account of this variability. This means, in mathematical parlance, treating them as probability distributions. Some modellers, particularly in the USA and Israel (Dagan and Bresler, 1979; Wagenet and Rao, 1983), have added this extra level of complexity to their already complex systems of differential equations and getting the whole structure to work is no mean achievement. Whether the results, in terms of simulating leaching in soils in the field are much better than those achieved with simpler models seems to remain an open question.

If would-be modellers of leaching decide that the very complex models are not satisfactory what are the alternatives? They can go back to the simpler functional models discussed earlier, or consider an entirely different approach to modelling.

Stochastic models for leaching presume that the soil varies so much that its variability is the dominant factor in leaching. This being so, effort devoted to the mechanisms of leaching is wasted; all that is needed is an ultra-simple description of the process backed by good statistics. The simplest description of the downward movement of solute is the travel time, the time the solute takes to move from the soil surface to a specified depth. This approach, developed in California (Jury, 1982), considers the probability that solute applied at the soil surface will arrive at a given depth following the application of a given amount of water. For the mathematically-minded

$$P_z(w) = \int_0^1 f_z(w) \, dw \tag{6.6}$$

where $f_z(w)$ represents the probability density function that summarizes the probability P_z that solute added at the surface will arrive at depth z as the quantity of water added at the surface increases from w to $(w + dw)$. The model treats the soil as an assembly of twisted capillaries of differing lengths through which the solute moves by piston flow. Calibrating the model for one depth and one amount of water enables it to predict the arrival of solute at a greater depth when further rain falls or more irrigation is applied. One problem for this approach may lie in soils that change in texture with increasing depth. It is also an approach that has been tested mainly with solute applied to the surface; whether it can treat the nitrate produced in the soil by microbes is not known.

We also do not know how well it will predict the flux of nitrate at a given depth, as opposed to predicting the arrival of a specific concentration, or for how long a calibration will remain valid. The simplicity of the model makes it a candidate for use as a management tool, but its usefulness in this way would be greatly lessened if it had to be recalibrated annually for each field.

Modelling the microbes: mineralization

Mineralization, the process by which organic nitrogen in soils breaks down to the inorganic form, can, like leaching, be modelled at several levels of complexity, but the nature of the complexity differs between the two processes. Making leaching models more complex involves making the description of the process more fundamental and the mathematics more sophisticated. Increasing the complexity of a model for mineralization usually means making it more detailed so that the computing problems are more organizational than mathematical. There is no clear distinction between mechanistic and functional models, like that for leaching models, mainly because there are almost limitless possibilities for making models of microbial processes more mechanistic. Should we, for example, stop at the kinetics of the reaction the microbes are mediating; or include the build-up and decline of the microbial population involved; or consider rival populations, microbial genetics or even molecular biology? There is also no distinction between *rate* and *capacity* models. All mineralization models consider the rates of conversion of one form of nitrogren to another and most of them involve 'pools' of nitrogen which tend to buffer the system against change in the way that a capacity parameter in a leaching model does. A 'pool' of nitrogen was defined by Dr David Jenkinson of Rothamsted as: 'a compartment of the soil nitrogen that is chemically indistinguishable and equally accessible to the plants or to the soil's population' (D.S. Jenkinson, personal communication).

Most models of mineralization are *deterministic*, but microbial processes have a strong element of the random, to us if not to the microbes; thus *stochastic* models are beginning to evolve but only to a limited extent so far. As before, the purpose of the model is an important consideration.

Some individual models for mineralization

Those who model mineralization have to cope with the problem that in most soils mineral nitrogen (both ammonium and nitrate forms) is not only released from the soil organic matter but also immobilized in it by microbes, with the balance between the processes determined by the amount of carbon available to the microbes. Measurements of mineral nitrogen in the soil show the net outcome of these two processes, which is usually an overall release of mineral nitrogen, but may be a net loss.

Simple models for net mineralization

If a sample of soil is incubated in a moist state at constant temperature the amount of mineral nitrogen in it usually increases steadily with time. Plotting the increase against time may give a straight line (e.g. Tabatabai and Al-Khafaji, 1980; Addiscott, 1983), which implies that the release of mineral nitrogen is independent of the amount of nitrogen available for mineralization. Another possibility is that a better straight line is obtained by plotting the increase against the square root of time. Alternatively, we may need to assume that the rate of release is proportional to the amount of potentially mineralizable nitrogen, N_m.

$$\frac{dN_m}{dt} = k_1 N_m \tag{6.7}$$

Integrating this relationship and assuming that $N_m = N_0$ when $t = 0$, and that N_t is the amount of mineral nitrogen released up to time t gives

$$N_t = N_0(1 - e^{-k_1 t}) \tag{6.8}$$

where k_1 is the rate constant (Stanford and Smith, 1972). This type of relationship is known to chemists as a 'first-order' kinetic relationship. All three relationships are empirical – that is to say they are based on experimental results rather than theory – and all three have been found to happen, so there is no basis for saying that one is better than the others. Which one occurs may depend on the way the soil is treated before the experiment. The simple linear relationship with time is most likely to be found if the soil is kept moist when it is brought in from the field. If the soil is dried and then re-wetted the first-order relationship will probably appear, but this may be an experimental artefact. Drying the soil and re-wetting it tends to stimulate a 'flush' or sudden release of nitrogen which could make the simple linear relationship look like a first-order one (Fig. 6.6). The relationship with the square root of time is very difficult to distinguish from the first-order relationship. All three relationships are over-simplifications of something that is more complex in reality, and there are strong arguments for using the linear one, which is the simplest, if there is nothing to suggest that one of the others would be better.

Mineralization proceeds faster as the soil becomes warmer and moister – up to a point. We therefore need to allow for the soil's temperature and water content. There are two main ways of simulating the effect of temperature on the rate of mineralization, both derive from the work of the physical chemist Arrhenius who investigated much of the kinetics of chemical reactions. The simpler is to use a factor known as the Q_{10}, which is the proportional increase in the rate that results from a 10°C rise in temperature. The Q_{10} for many biological processes is about 2, which means that

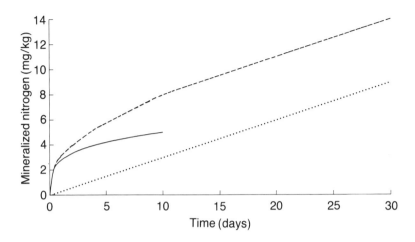

Fig. 6.6. Why an apparent first-order release of nitrogen (------) may be an artefact. If the soil has been dried, re-wetting it may cause a sudden rapid 'flush' of mineralization (——), which if added to a linear (zero-order) release of nitrogen (....) could give the apparent first-order release.

the rate roughly doubles for each 10°C rise. The other is to use the relationship derived by Arrhenius himself,

$$Rate = Ae^{-\frac{B}{T}} \tag{6.9}$$

where T is the absolute temperature and A and B are constants. The constant B is about 7000 K for mineralization (Addiscott, 1983) and it is usually easier to avoid the use of A by defining the rate at a specified temperature. If the soil becomes too hot (more than about 35°C) the microbes cease to function and they become less active as the soil approaches freezing point. They also do not like to be too wet or too dry; there is usually a plateau (Fig. 6.7) at which the rate of mineralization is independent of soil moisture. At the dry end of this there is a steady decline as the soil becomes drier, and there will probably also be a sharper decline as the soil passes a certain degree of wetness at the other end.

Models with two or more 'pools' of mineralizable nitrogen

Ever since agriculture began, farmers have appreciated the benefits to be derived from growing crops in rotation, that is growing a different crop every year. This practice helps to control disease, but it can also supply nutrients because some plants, such as legumes, put in more nitrogen than

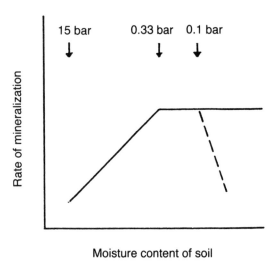

Moisture content of soil

Fig. 6.7. Effect of changing water content on the rate of mineralization. The rate declines when the soil is too dry and may, but not always, decline sharply when it is too wet. (Based on Stanford and Epstein, 1974.)

they take out. A model, even a simple model of mineralization, needs to account for differences in the power of differing categories of soil and crop residues to supply nutrients. The simplest possibility lies in a two compartment model. One compartment breaks down relatively rapidly while the other turns over much more slowly. There is usually much more material *potentially* available for decomposition in the more resistant pool because the more rapidly decomposing material comes only from recent crop residues. At first then the contribution to mineralization from both pools may be about the same, but as the microbes consume the easily decomposed material the resistant pool becomes dominant. Some authors represent this scheme by treating the resistant pool as too large to be changed appreciably by the release of nitrogen from it so that the release has zero-order kinetics. In mathematical terms the process is described by an equation with both an exponential and linear term:

$$N = N_e(1 - e^{-k_e t}) + N_r k_r t \tag{6.10}$$

where N_e is the organic nitrogen in soil that is more easily broken down and k_e the rate at which it decomposes, and N_r the pool containing nitrogen that is more resistant to change with rate constant k_r. Other models treat both pools as finite so that their decomposition is described by a double exponential decline:

$$N = N_e(1 - e^{-k_e t}) + N_r(1 - e^{-k_r t}) \tag{6.11}$$

More sophisticated models contain even more pools. Most include a description of the living soil microbial biomass but some use more complicated kinetics to describe the build-up and death of these organisms. Most include a resistant fraction known as the soil humus; often they partition the humus into material that is very highly resistant to chemical attack by micro-organisms and some that is physically protected from attack because it is held in the tiny gaps between clay surfaces into which the microbes cannot penetrate. Others partition the living organisms between large or small, animal or plant, predator or prey. The astute reader will realize that it is often the scientist's interest that determines the nature of the model. But even the most mechanistically sophisticated model approximates the many genera of micro- and macro-organisms in soil and the way in which these organisms distribute themselves in three dimensions. The models are best classified according to the job they have to do. Many multi-compartment models have helped our understanding of soil processes greatly by shedding light on possible flows of nitrogen through soil organic matter. It is virtually impossible to measure these changes in any way. On the other hand the level of complexity of these models is such that they require very many parameters to be estimated and they are unlikely to be of much help to scientists who need to estimate, for example, the contribution of large areas of agricultural land to nitrate concentrations in aquifers. In this case the simpler models with one or two pools, but fewer parameters, have a much greater chance of being reliable.

Models of crop growth and integrated models of the nitrogen cycle

Mineralization is of importance to scientists seeking to model leaching because of the importance of the rate of turnover of organic matter in soil in determining how much nitrate leaches. For much the same reason, models of crop growth and nitrogen uptake are important to modellers of leaching because they determine how much nitrogen is immobilized in crop residues, or removed from the land in the crop, and how much remains at risk. A detailed discussion of the workings of crop models is outside the scope of this book but the interested reader will find much to help him in a book by France and Thornley (1984).

The way in which models of the different parts of the nitrogen cycle can be integrated might seem obvious; all that is needed is to keep track of where nitrate is in soil, how deep and how effective plant roots are and at what rate soil organic matter decomposes. In practice it is not always so simple. Imagine the difficulty in solving Equations (6.4) and (6.5) for every point, for every time interval and for several depths within the soil, possibly

using computer intensive numerical methods. Then imagine the complexity
of solving equations for the break-down or synthesis of organic matter and
the activity and growth of the plant – integrating the modelling of all three
processes! All these processes interact. Organic matter breaks down more
slowly when the soil is dry; this will happen when the crop extracts water so
interrupting the downward flow of nitrate in the process. Numerical
solutions to the equations describing leaching may involve iteration, so that
to use truly mechanistic equations to describe agriculture at the field scale
requires enormous amounts of computer time, so much in fact that there
are occasional suggestions that it might be quicker to wait and see what
happens in nature!

For these processes, simpler, functional models are advantageous.
Addiscott and Whitmore (1987) took a capacity model of water and
nitrate flow, a model of net mineralization of nitrogen using zero-order
kinetics and a simple function describing the uptake of nitrogen from soil.
Their model describes the movement of nitrate under arable crops during
winter reasonably well (Fig. 6.8). Greenwood *et al.* (1987) have produced

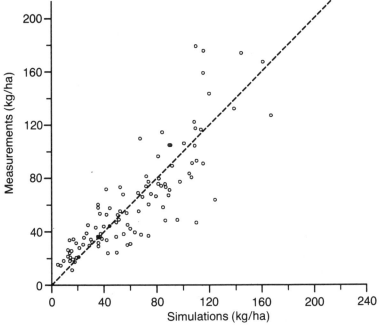

Fig. 6.8. Using models for leaching, mineralization and crop uptake together.
Comparison of measured and simulated values for the quantity of mineral nitrogn
(NH_4 plus NO_3) in the soil to 0.9 m. The relationship between the measured and
simulated quantities of nitrogen in the crop was better than this but the
corresponding relationship for dry-matter production not so good. Data from seven
sites during 5 years.

a similar model for wheat based partly on Burns' leaching model. Variants of it, with simple but effective descriptions of mineralization, simulate leaching under vegetables and potatoes. Richter *et al.* (1985) has described mineralization in arable soils in Germany with two pools as in Equation (6.10). Hansen and Aslyng (1984) in Denmark worked along similar lines but had the courage to cover denitrification which is not usually found in simple models such as these. Thornley and Verberne (1989) tackled the immense problem of nitrogen cycling in grazed grassland but paid little attention to leaching because grass is often grown on heavy land that is impervious to water and so unsuitable for arable crops. Molina *et al.* (1983) in the USA, and McGill *et al.* (1981) in Canada have all attempted something along the same lines as these European modellers. This survey is not intended to be exhaustive, but is simply intended to give the interested reader an introduction to the literature on the subject.

No model is perfect; if it were it would cease to be a model. Models are useful, not because they reproduce the reality, but because they simplify it and enable the most important aspects to be identified, studied, simulated and ultimately predicted in advance. It is unwise to claim too much for models of nitrate leaching or any other processes, but if they are used with caution they can enhance the value of experimental data and suggest new approaches to decreasing nitrate leaching from land.

Chapter Seven
Sources of Nitrate Leakage: Arable Farming

Natural systems of vegetation do not waste their resources. They have usually evolved to occupy a particular niche in the most efficient way possible, and there are other species able to replace those that become profligate with the soil's bounty. Well-established grassland or forest loses very little nitrate to the rest of the environment. There is a continuous annual cycle of growth, senescence and decay which includes the production of nitrate from dead leaves and roots, but this nitrate can be repossessed by the roots of the grass or the trees before it escapes to become a problem (see also Chapter 4). Measurements of nitrate in water draining from mature grassland or forest usually show small nitrate concentrations.

Arable farming is far from being a natural system. This is not to say that it is a bad system – the world would be much hungrier without it – and it is certainly a very old system. One of Christ's most noted parables concerned an arable farmer who went out to sow, with rather variable results. And arable farming can be found far earlier than his time, both in the Old Testament and in archaeological records. We have to accept, however, that arable farming contributes to the nitrate problem; to understand why it does so, so that we can limit the problem, it is necessary to consider the ways in which arable farming differs from a more natural system of vegetation.

One crucial difference separates arable farming from grassland, forest or other systems of vegetation. No other system ever leaves the soil bare of vegetation; arable farming, by its very nature, has to leave the soil bare for part of the year, and during this period there are no living roots to capture any nitrate that is in the soil. Figure 7.1 shows how sowing winter wheat early lessened the amount of nitrate in the soil during autumn and winter and thereby lessened the risk of leaching. The length of time for which the soil is bare depends greatly on the sequence of crops. In the UK, potatoes or sugar beet, which are sown in spring, may follow a winter wheat crop harvested at the end of July, leaving the soil bare for as much as 8 months.

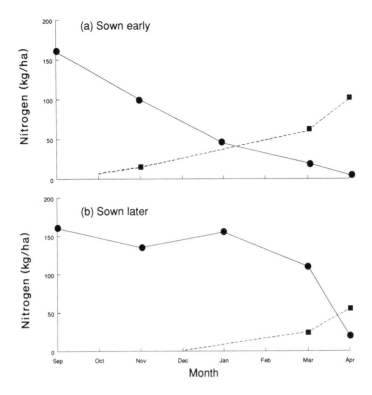

Fig. 7.1. Sowing winter wheat early increases the uptake of nitrogen by the crop and lessens the amount of mineral nitrogen at risk in the soil in autumn and winter. Soil nitrogen (—●—); crop nitrogen (--■--) (Widdowson *et al.*, 1987).

By contrast, an early-sown crop of winter oilseed rape could be in the soil less than a month after the wheat was harvested. The gap can be even smaller in some systems of tropical agriculture. Temperatures are high and fertile land is in short supply, so three crops may be grown in a year.

A second factor differentiates most arable farming from other systems, and that is cultivation. The land is tilled to control weeds and to provide a favourable seed-bed, and this tends to stimulate microbes in the soil to produce nitrate. Thirty years ago it was not uncommon for farmers to plough the soil in autumn while it was still reasonably dry and then leave it bare over the winter ready for a crop in spring. This meant that nitrate production was stimulated but that there were no active roots to capture it. The consequences can be seen in the sharp increase during autumn in the quantity of nitrate carried by the River Thames (Fig. 7.2). This increase was more marked in the early 1970s than the early 1950s, probably for two reasons.

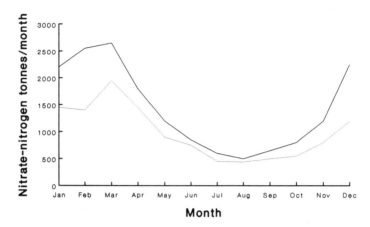

Fig. 7.2. Changes during the year in the quantity of nitrate carried by the River Thames 1950–54 (----); 1970–74 (–) (taken from Jenkinson, 1986).

1. The fertility of the soils in the catchment may have increased overall so that more nitrogen was mineralized during autumn.

2. Farmers were still applying some nitrogen fertilizer to crops in the autumn and much of this nitrogen was wasted because it was washed out of the soil; if autumn applications followed the upward trend of other applications of nitrogen fertilizer during this period they would have contributed to the increase, perhaps disproportionately.

Far, far less soil is now ploughed and then left bare over winter, and the use of nitrogen fertilizer in autumn has declined markedly, so the autumn increase in the nitrate load of the Thames should now be less than it was. Figure 1.1(c) shows that this was indeed so in the early 1980s although the decline was small; the effect was rather more marked in other rivers, for example, the Lee.

In agriculture it is convenient to think in terms of two periods of the year rather than the customary four seasons. The first of these periods is when the crop is growing vigorously and actively taking up nitrate and other nutrients through its roots, and the other is when it is not. The farmer, the crop and the microbes in the soil are all influenced by these two periods, so it is logical to give them separate consideration.

The active growth period

The onset of active growth and nitrate uptake is marked in different ways in different parts of the world: it may be spring thaw or the arrival of the rains, but it is often clearly definable. In the UK and climatically similar countries, however, the vagaries of the weather make the event less obvious, but a useful working definition might be the time at which the lawn first needs to be mown regularly. The lawn, like nearby crops, is responding to increases in the temperatures of the soil and air, longer day lengths and more intense sunlight. Photosynthesis increases and crop growth accelerates, necessitating the production of proteins from mineral nitrogen (nitrate and ammonium) taken up from the soil, and root growth accelerates too. Soil microbes also respond to the warmth and increase their production of ammonium and nitrate. Unfortunately, the united efforts of the roots and the microbes rarely supply the quantity of mineral nitrogen needed by the crop, and more has to be supplied if the growth of the crop is not to be held back (Chapter 2).

The extra nitrogen can be supplied by organic farming methods (Chapter 8), but most farmers in the UK and other temperate regions supply it as chemical fertilizer. A large proportion of farmers in tropical areas do the same. The different kinds of nitrogen fertilizer and their behaviour in soil were discussed earlier (Chapters 2 and 3) and the key point to recall is that all the forms of nitrogen fertilizer are likely to be converted to nitrate by microbes in the soil. Just where the nitrate goes is, needless to say, one of the central questions to be discussed in this book. There are four main possibilities (Fig. 7.3).

1. It may be absorbed by the crop, as the farmer intended.
2. It may be incorporated in the soil's organic matter, where it will cause no problems until the microbes re-mobilize it.

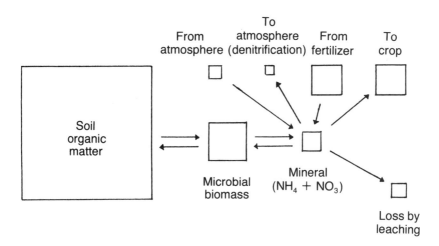

Fig. 7.3. Nitrogen pathways in soil. The quantities of nitrogen likely to be in each form are proportional to the areas of the squares.

3. It may be washed out of the soil and potentially become part of the nitrate problem.
4. It may be denitrified. This means that the microbes in the soil use some of the oxygen (O) so that nitrate NO_3^- becomes dinitrogen (N_2) or nitrous oxide (N_2O); both are gases.

The first of these outcomes is desirable, the second satisfactory and the third obviously highly undesirable. If the gas formed in the fourth outcome is dinitrogen, the result is a waste of the nitrogen fertilizer and the energy used to produce it but no more. The production of nitrous oxide, however, is more than a waste; it is as much of an environmental problem as the leakage of nitrate from the soil, as we saw in Chapter 4. The longer the nitrate remains in the soil, the more likely it is to be absorbed by the crop or another plant or to become incorporated in organic matter. But, as we shall see, the more rain falls after nitrate is applied, the more likely it is that the nitrate will either be washed out of the soil or be denitrified.

It is obviously very important to find out where nitrogen applied as fertilizer ultimately goes; how much is harvested with the crop, how much stays in the soil in organic matter, and how much is either washed out or denitrified. Unfortunately, it is also very difficult to find this out. Measuring the quantity of nitrogen in grain and straw tells you how much was removed from the soil but not whether it came from fertilizer or from mineral nitrogen released by microbes in the soil. Measuring the amount of fertilizer nitrogen that stays in the organic matter would involve measurements of the nitrogen in organic matter before and after the crop was grown and the nitrogen fertilizer applied. Both the 'before' and 'after' measurements would show that the soil contained about 25 times as much nitrogen in organic matter as was applied as fertilizer. The difference between the two measurements will be very small and will not be a trustworthy guide to the amount of nitrogen from fertilizer left in the organic matter. This approach is, to borrow an illustration from David Jenkinson, like weighing a ship's captain by weighing the ship and then weighing it again with the captain aboard – and getting his weight by difference. Direct measurements of losses of nitrate from the soil are not very easy or reliable in some soils (Chapter 5), and they do not show whether the nitrate that is measured comes from the soil or the fertilizer. Measuring denitrification reliably is even more difficult than measuring leaching; the technical difficulties can be almost insuperable in some soils. Add to this *mélange* of problems the mineral nitrogen deposited from the atmosphere (Chapter 9) and the atmospheric dinitrogen captured by organisms in the soil (Chapter 4) and it becomes clear that tracing the fate of nitrogen fertilizer by sets of conventional measurements is just not practicable.

Tracing the fate of nitrogen applied as fertilizer

The 'tagging' of birds and animals to find out where they go is a concept familiar to most people. A similar approach is used to trace the fate of nitrogen fertilizer. The 'tag' or 'label' is the heavy isotope of nitrogen, ^{15}N. This behaves in almost exactly the same way as the ^{14}N normally encountered, and ^{15}N used to label fertilizer nitrogen is absorbed by the crop, retained in the soil's organic matter, leached or denitrified along with the rest of the fertilizer nitrogen. The ratio of ^{15}N to ^{14}N in the fertilizer is known, so when the crop and the soil are analysed for total nitrogen and then for their ^{15}N to ^{14}N ratio, the amount of nitrogen in the crop and the soil that came from the fertilizer can be calculated. One point needs to be mentioned; it is sometimes thought, wrongly, that ^{15}N is a radioactive isotope. It is not. If it was, it would be much easier to detect but it would have to be used with much more caution (and considerably more paperwork). The only way of discriminating between ^{15}N and ^{14}N is by their atomic masses, and a mass spectrometer has to be used to determine the relative amounts of the two isotopes. The overall analytical procedure is neither easy nor quick; the total amount of nitrogen in the sample has to be measured by conventional chemical means and then a further sample has to be brought into the mass spectrometer for the ratio of ^{15}N to ^{14}N to be determined.

This approach is very demanding in terms of time, equipment and skilled manpower, but the investment of these resources has proved well worthwhile. Research with ^{15}N throughout the world has provided insights of critical importance to the understanding of the nitrate problem (e.g. Smith *et al.*, 1984; Powlson *et al.*, 1986; Recous *et al.*, 1988). One example is the research at Rothamsted in which nitrogen fertilizer labelled with ^{15}N was applied to winter wheat in field plots on three differing soil types, flinty silt loam, sandy loam and sandy clay over a heavy clay subsoil. Fifty to 80% of the labelled nitrogen was found in the crop and a further 10–25% was in the soil when the crop was harvested (Fig. 7.4; Powlson *et al.*, 1991), but – this was the key finding – almost all this nitrogen in the soil was in *organic* forms. It was in roots, microbes or the various other types of organic matter in soil. In most cases only about 2% of the nitrogen from fertilizer was left in the soil as nitrate (for example, Fig. 7.5 taken from MacDonald *et al.*, 1989). Nearly all the nitrate found was *not* labelled and therefore did not come from the fertilizer. The amount of this unlabelled nitrate bore no relation to the quantity of fertilizer supplied. Thus the idea that the nitrate problem was caused by nitrate fertilizer left unused in the soil by the crop seemed to be refuted. This conclusion will be discussed further, later in the chapter, in the light of information from crops that leave more unused nitrate from fertilizers than winter cereals do.

Although the experimenters could account for some very satisfactory

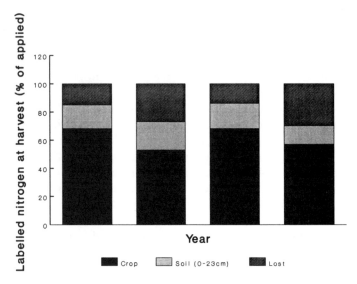

Fig. 7.4. The fate of ^{15}N-labelled (tagged) nitrogen applied as fertilizer to winter wheat (Powlson *et al.*, 1992).

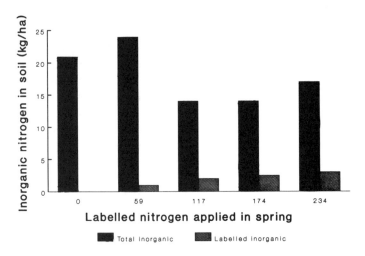

Fig. 7.5. Quantities of labelled and unlabelled inorganic nitrogen in soil at harvest following various spring applications of ^{15}N-labelled fertilizer. (MacDonald *et al.*, 1989). This inorganic nitrogen was mainly nitrate.

percentages of the labelled nitrogen in the crop and in the soil, some always remained unaccounted for and had to be presumed lost. The percentages ranged from 8 to 35% with an average of 15% and these, we should stress, were losses that occurred between the time that the fertilizer was applied and the time that the crop was harvested. How did they occur? The losses were usually much greater if the application of nitrogen fertilizer was followed by above average rainfall, and Figure 7.6 shows that they could be related best to the amount of rain falling during the first 3 weeks after the fertilizer was applied (Powlson *et al.*, 1992). This relationship provided a useful guide to likely losses in future, but since rain encourages both leaching and denitrification, it gave no indication as to which process was responsible for the losses from these experiments. Partitioning such losses between the two processes when both are likely to occur is not easy; both are difficult to measure in the field.

The partitioning was attempted for this particular set of experiments using computer models to estimate the percentage of the labelled nitrogen that was lost by leaching (Addiscott and Powlson, 1992). This involved using the SLIM leaching model (Chapter 6) and also a model for uptake of nitrogen, because the crop absorbed some of the labelled nitrogen, and a mineralization model (Addiscott and Whitmore, 1987) because the crop took up nitrogen produced by microbes in the soil as well as that from the fertilizer. The resulting estimates of the percentage losses of labelled

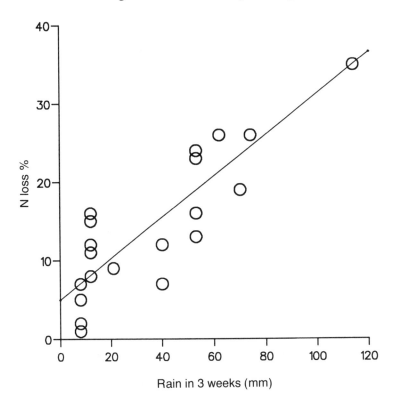

Fig. 7.6. The relationship between the loss of nitrogen fertilizer labelled with ¹⁵N and the amount of rain in the 3 weeks after application.

nitrogen by leaching were subtracted from the total percentage losses to provide estimates of the losses by denitrification. These apparent denitrification losses could have resulted simply from an underestimate of leaching, but a check that was made suggested that this was not the case. Assuming, on the basis of this check, that the labelled nitrogen apparently lost by denitrification had indeed been lost in this way the total loss of labelled nitrogen could be partitioned between the two pathways of loss (Fig. 7.7). In two of the 13 experiments examined the loss was totally by leaching, although one of these losses was very small. In another experiment the loss was totally by denitrification, but in the remaining 10 experiments both pathways contributed. Overall, however, nearly two-thirds of the total loss of labelled nitrogen was by denitrification. This meant that of the average loss of 15%, leaching contributed only about 5%. As only 1% of the labelled nitrogen was left in the soil as nitrate at harvest, direct losses of nitrate from the labelled fertilizer nitrogen could have been no more than about 6%. These carefully-made scientific experiments lent no support at all to the idea widely spread by the media that the nitrate problem resulted from massive losses of nitrate from fertilizer wantonly spread by greedy farmers. That being said, however, some cautionary notes need to be added.

One is that these experiments were made with more attention to detail than is possible on ordinary commercial farms. In particular, the fertilizer nitrogen was applied more evenly than would be possible on a farm scale.

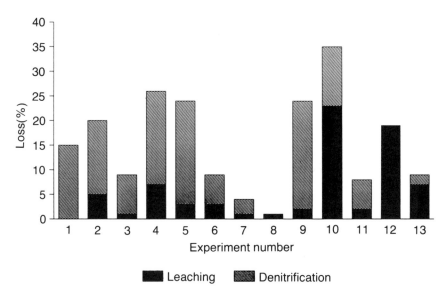

Fig. 7.7. The partitioning of the losses of ^{15}N-labelled fertilizer between denitrification and leaching. These losses occurred between the application of the fertilizer in spring and the harvest of the crop (Addiscott and Powlson, 1992).

Direct losses by leaching may be a little larger on commercial farms, especially if the fertilizer spreader is not set accurately or if it passes with the spreader overlapping.

Another point is that these experiments were made with winter wheat, a crop that already has a well-established root system when fertilizer is applied. Should we draw conclusions from this one crop? Part of the answer to this question is 'yes', simply because about 60% of arable land in England grows winter cereal crops. Winter wheat is also reasonably representative of other crops that are sown in the autumn and establish their root systems before spring. Oilseed rape, for example, was not found to lose very much labelled nitrogen before harvest or to leave much in the soil at harvest. It does, however, leave residues in the soil that cause problems later in the season, and these are discussed elsewhere in this chapter. Other crops do suffer large losses of nitrogen fertilizer. Potatoes need fairly large amounts of nitrogen but they are not efficient at taking it up from the soil, partly because their root systems are not as extensive as those of cereal crops. Another problem with potatoes is that the fertilizer is often applied at the time that the crop is planted so that nitrogen fertilizer may be leached even before the crop has really started to grow. Figure 7.8 shows the amounts of mineral nitrogen left in soil at harvest by wheat, rape and potatoes (MacDonald *et al.*, 1990). It also shows how much labelled nitrogen (that is from fertilizer) was found in soil at the same time. It is clear that potatoes present the greatest risk and that oilseed rape is marginally more risky than winter wheat. Note that the amount of mineral

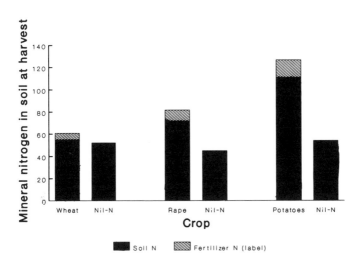

Fig. 7.8. The quantities of labelled and unlabelled mineral nitrogen in the soil after the harvest of winter wheat, rape and potatoes. Nil-N is no nitrogen application.

nitrogen in soil in the plots receiving no nitrogen fertilizer were the same for all three crops.

The conclusion that on average only about 6% of the labelled fertilizer nitrogen applied to winter wheat was lost directly by leaching is very important for the overall understanding of the nitrate problem. It depended in part, however, on a computer modelling exercise, and some may prefer to see more direct evidence that direct losses from fertilizer are as little as this. Lysimeters and the Brimstone experiment (Chapter 6) have provided such evidence. When spring barley was grown in a lysimeter on shallow soil overlying chalk (Dowdell *et al.*, 1984), only 6.3–6.6% of the labelled fertilizer applied to it was found as nitrate in water draining from the soil, despite the fact that the researchers collected the drainage for another 3 years after the labelled application. There are no directly comparable data from the Brimstone experiment, but nitrate losses in the drainage after nitrogen fertilizer was applied in the spring up to the cessation of drainage were mainly less than 10 kg ha^{-1}, again probably corresponding to roughly 6 or 7% of the fertilizer applied (Goss *et al.*, 1988). Thus neither experiment suggested that the direct loss of fertilizer nitrogen by leaching is likely to be very much greater than the 6% deduced above.

This satisfactory state of affairs depends on two factors: the removal of nitrate from the risk of leaching by the crop and the removal of the risk as the soil dries out, a process accelerated by the crop's withdrawal of water from the soil. When the crop is harvested the first of these safety factors is removed and the second lessened in its effect, and harvest signals a definite change to the second of the two periods of the year to which we referred at the start of the chapter.

After harvest

Harvest is the point at which arable land becomes completely different from grassland and natural systems of vegetation; the soil is made bare and then usually cultivated soon afterwards. This sharp difference between arable farming and natural systems of vegetation is one reason for looking for the roots of the nitrate problem in the period after harvest. Other reasons emerge as we think about what happens to water and nitrate in the soil during this period.

In the UK, winter may feel wetter than the summer, but this is not usually because more rain falls. The average daily rainfall varies very little between months (Fig. 7.9), but there is a change in the amount of evaporation. During autumn the amount of radiation reaching the land surface each day becomes smaller as days become shorter and the sun's radiation more attenuated, so there is less energy to evaporate water. Strong

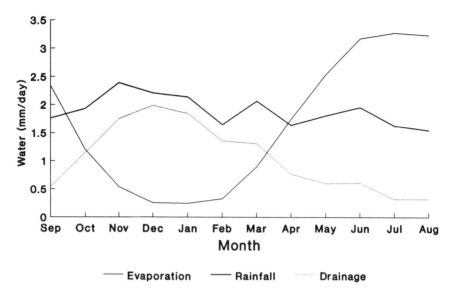

Fig. 7.9. Variations during the year in the average amounts of rainfall, evaporation and drainage through 0.5 m soil expressed on a daily basis.

evaporation during summer usually leaves the soil fairly dry at harvest, but during the autumn rainfall overtakes evaporation and the soil becomes moister. Eventually water begins to flow downwards through the soil carrying with it any nitrate that is there. It usually continues to flow until the spring. When the soil re-wets in autumn, therefore, it is vital to have as little nitrate in it as possible. Any nitrate left from fertilizer will almost certainly contribute to pollution because there will either be no crop growing or a crop that is not yet taking up much nitrate; the longer the gap between crops the greater the risk, so spring-sown crops increase the risk of pollution unless a 'catch-crop' is grown in the intervening period to pick up the nitrate (see Chapter 10).

We saw earlier in this chapter that only about 1% of the nitrogen fertilizer applied to winter wheat crops remains in the soil as nitrate at harvest and that the total loss of nitrate directly from the fertilizer is probably of the order of 6%. Six per cent of an average 190 kg ha^{-1} application of nitrogen is 11 kg ha^{-1}. Far more is lost in practice; so where does it all come from? Chapter 4 described how microbes in the soil are involved in the cycling of nitrogen and, in particular, how they break down nitrogen-containing organic material and release nitrate. The microbes, unfortunately, release nitrate when soil conditions suit them rather than when it suits the crop. Autumn, when the soil is still warm and becoming moister, suits them very well and they produce nitrate vigorously. This nitrate is vulnerable to leaching for the same reason as any nitrate left from fertilizer is vulnerable; water is starting to flow through the soil and there is little, if any, crop to catch the nitrate. We believe that this naturally occurring nitrate is the biggest cause of the nitrate problem – responsible for a far greater proportion of the nitrate problem than direct losses from nitrogen fertilizer.

This important recent conclusion is supported by the results of a very old experiment at Rothamsted. We referred in Chapter 5 to the Drain Gauges constructed at Rothamsted by J.B. Lawes and J.H. Gilbert in 1870. These gauges not only provided the first evidence of mobile and immobile water in soil, they also provided some invaluable information on nitrate losses from organic matter in the soil (Addiscott, 1988). From 1877 to 1915 the nitrate concentration was measured in each day's drainage. During the first 7 years of these experiments the soil leaked an average of 45 kg N ha^{-1} year^{-1} as nitrate. The soil in the gauges carried no crop and received only enough cultivation to remove weeds. It had received no fertilizer since 1868, so those 7 years were the 9th to 16th years since the last application. Only 3–5 kg of nitrogen was brought in by rain each year; almost all of the remaining 40–42 kg must have come from organic matter. The quantity of nitrogen leaked from the organic matter as nitrate in drainage during the whole 38 years of the measurements, about 1450 kg ha^{-1}, was almost exactly the same as the measured decline in the amount in the soil of the gauges, about 1410 kg ha^{-1}. Mineral nitrogen in the soil is a minute fraction of the total, so practically all the losses must have come from the soil's organic matter.

Another key point to emerge from these old data was that nitrate leakage continued for a very long time (Fig. 7.10). Calculations showed

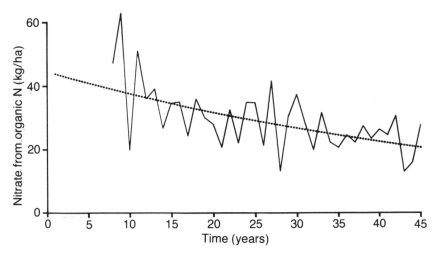

Fig. 7.10. The leakage of nitrate derived from nitrogen in organic matter in the 1.5 m Drain Gauge at Rothamsted, 1877–1915. The broken line is the result of removing the effect of fluctuations in annual rainfall (Addiscott, 1988).

that it took 41 years to decline to half its initial rate and that it would have taken well over a century to fall to one-tenth of it. There was obviously a large pool of decomposable nitrogen that was broken down very slowly, and this slow and inexorable loss seems unlikely to be affected by minor changes to arable farming systems. The Drain Gauges were also no respecters of European Community (EC) legislation. The water draining from this unfertilized soil in the 1870s initially had a nitrate–nitrogen concentration that exceeded today's EC limit of 11.3 mg l^{-1}. It took about 7 years before the concentration fell to within the EC limit.

We have seen that the amount of nitrate lost directly from nitrogen fertilizer by leaching is small and that microbes produce substantial quantities of nitrate from organic matter, some of it at the time of peak risk. Does this mean that fertilizers are not much to blame for the nitrate problem? Before we can answer this question we need to find whether there are any 'memory effects'. Is the amount of nitrate produced by the microbes in autumn influenced by the quantity of nitrogen fertilizer given to the crop that has just been harvested? Has the fertilizer perhaps increased the nitrogenous residues that microbes can break down easily? This is a very important question. The surplus nitrate curve discussed in Chapter 2 tends to show that increasing the amount of nitrogen fertilizer given to winter crops has no effect on the amount of nitrate left in the soil at harvest until a substantial amount has been given. There is a 'plateau' in the amount of nitrate left that can extend over most of the range of nitrogen applications that a prudent farmer would normally consider. This apparently safe plateau completely loses its relevance if there are memory effects that imply that extra nitrogen fertilizer means extra nitrate produced by the microbes after harvest. Do such memory effects occur? The answer seems to depend on the crop and the number of years for which it is grown at the same site.

Research at Rothamsted suggests that winter wheat grown for 1 year does not leave a memory effect; the rate at which microbes produce nitrate seems to be unaffected by the amount of nitrogen fertilizer supplied. Winter wheat has been grown continuously for 147 years in the Broadbalk Experiment at Rothamsted, and, perhaps not surprisingly, there is a definite memory effect (Fig. 7.11). The rate of production of nitrate by the microbes increases clearly with the amount of fertilizer that has been given each year. We are obviously keen to find out how long such a memory effect takes to develop. With some crops, unfortunately, the memory effect seems to occur after the first crop.

Oilseed rape seems to leave residues that break down to a greater extent than those of wheat in the months after harvest. This is possibly because of its habit of shedding its leaves while the pods are ripening, which means that the shed leaves are available to the soil's microbes even before the crop is harvested. It may be for this reason that oilseed rape is

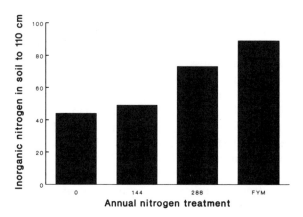

Fig. 7.11. Memory effect. Quantity of mineral nitrogen in the soil in autumn in plots given 0, 144 or 288 kg ha^{-1} of fertilizer nitrogen or 35 t ha^{-1} of farmyard manure (FYM).

the crop that has shown the most clear evidence of a memory effect after the first crop. Researchers in the Agricultural Development and Advisory Service found that nitrate production by microbes after a rape crop had been harvested increased with the amount of nitrogen fertilizer that had been given to the crop (R. Sylvester-Bradley, personal communication). With oilseed rape, then, the amount of nitrate left in the soil at harvest may not be a reliable guide to the after-effects of the fertilizer. It will be important to know how many other crops leave memory effects.

It is not only soil microbes that respond to these memory effects; if these effects mean more nitrate in the soil in autumn, more will be taken up by crops sown at this time. To evaluate fully these memory effects we need to take account of the extra uptake of nitrate by the crop as well as the extra production by the microbes in the soil. This is not easy. We referred above to the memory effect in the Broadbalk experiment at Rothamsted; here we knew for each of the standard annual applications of nitrogen how much nitrate was left in the soil at harvest and how much nitrogen was taken up by the crop during the year. We also know how much effect the long term applications of nitrogen had on the rate of production of nitrate by microbes. All this information was supplied to a computer model which used it to provide estimates of the concentration of nitrate in water leaving the root zone (Fig. 7.12). It appeared that giving a little nitrogen, 48 kg ha^{-1}, increased this nitrate concentration relative to that in water leaving the unfertilized plot but that further additions up to 192 kg ha^{-1} brought no further increase; the concentration remained on a plateau. The 192 kg application was the largest given but we suspect that applications much larger than this would have increased the concentration above the level of

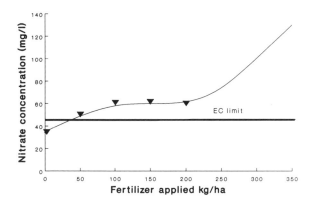

Fig. 7.12. Estimates of the nitrate concentration in the water draining from the plots of the Broadbalk experiment, obtained by applying a computer model to the measured amounts of nitrate in the soil in the autumn, amounts of nitrogen taken up by the crop and the measured rates of mineralization. The shape of the curve beyond the points is based on the assumption that less and less of the nitrogen applied beyond 192 kg ha^{-1} is taken up.

the plateau. Larger applications, 240 and 288 kg ha^{-1} – well above the amount needed to maximize yield on Broadbalk – have been made during the past few years; the amounts of nitrate in the soil at harvest and during the following winter have been substantially larger than in plots given applications closer to standard practice. The shape in Figure 7.12 is very similar to the shape of the surplus nitrate curve in Figure 2.3, which is not surprising because nitrate left in the soil at harvest is very vulnerable to leaching. The quantity of nitrate in the soil at harvest time is not *necessarily* a reliable guide to the overall risk of nitrate leaching, because it will be augmented by nitrate produced later by mineralization, but some recent work at Rothamsted suggests that this later mineralization may not be a major problem. When a wide range of soils, which had carried various preceding crops, were ranked for nitrate content at harvest and in November the rankings did not change between the two samplings (A.J. MacDonald, unpublished data).

Memory effects are just one of the facets of the complex nitrogen cycle that was described in Chapter 4. These effects are important because we shall not be able to draw a firm conclusion about the role of nitrogen fertilizer in the nitrate problem until we understand them fully. This means, in effect, that we shall have to extend our understanding of the nitrogen cycle and improve our models of the turnover of organic nitrogen in soil. This is urgently needed.

A large proportion of arable land devoted to producing winter cereals –

62% in the UK – helps to restrict the nitrate problem. Other crops seem to leave not only more nitrate in the soil at harvest but also a larger percentage of the fertilizer nitrogen they receive (Fig. 7.8 above). Potatoes given a standard application of nitrogen (about 220 kg ha^{-1}) left twice as much nitrate in the soil as winter wheat, and this amount was also twice the amount left by potatoes given no fertilizer; that is, there was an immediate marked memory effect with potatoes. The amount left by oilseed rape lay between the amounts left by winter wheat and potatoes. Some caution is clearly needed when drawing conclusions from experiments with ^{15}N. With winter wheat it is correct to conclude that, of the 60 kg ha^{-1} of nitrogen left as nitrate in the soil, only about 5 kg ha^{-1} came from fertilizer and decreasing the fertilizer application below the standard would have made practically no difference to the amount of nitrate left in the soil. When, however, potatoes were given ^{15}N labelled fertilizer, not only were 16 kg ha^{-1} of the 120 kg N ha^{-1} of nitrogen left as nitrate in the soil labelled and therefore from the fertilizer, but the difference in nitrate-nitrogen left between crops with and without nitrogen fertilizer was about 60 kg ha^{-1}, not 16 kg. This disparity probably occurred because the nitrate from the fertilizer was more readily available to the potato crop than the indigenous soil nitrate; the fertilizer was probably closer to the roots when needed. It shows how unwise it is to try to extrapolate the results of experiments from one crop to another.

Arable farming, and nitrogen fertilizers in particular have been very much 'in the dock' in the past few years in the UK and in other countries with similar climates and agriculture. We can perhaps conclude this chapter most usefully by summarizing the cases for the 'prosecution' and the 'defence'.

The case against arable farming

During the past few decades the production of arable crops has increased dramatically to the extent that Europe has seen a surplus of grain and other commodities. This increase has been achieved, at least in part, by increased use of nitrogen fertilizer. During this time the concentration of nitrate in natural waters has increased sharply, to the potential detriment of the community at large. The prosecution notes that farmers' incomes have also increased sharply during this period and feel that this has been achieved by farming methods that are too far removed from a natural system of vegetation.

The case for the defence

Production is not a crime. Without it the world's population would have
nothing to eat, wear or use. The quantity of any commodity that is
produced is set by the consumer as well as the producer, and politicians
may have an over-riding influence, as we have seen with the Common
Agricultural Policy. As recently as 1975 a White Paper (statement of
policy) produced by the British Government stated that 'The government
takes the view that a continuing expansion of food production in Britain
will be in the national interest' (HMSO, 1975). Any current surplus of
grain in Europe is much preferable to the famine in the years immediately
after the war, and there is certainly not a surplus of food in the world as a
whole. Nitrogen fertilizers have indeed played an important role in the
increased production of arable crops. Without them the increase would not
have been possible and the soil's fertility might have decreased as more
nitrogen was taken from it than was returned. If nitrogen fertilizers have
contributed to the nitrate problem it is not necessarily because they have
been used to excess. Sensible applications of nitrogen fertilizer do not leave
large amounts of nitrate in the soil, except with minority crops, such as
potatoes, that do not occupy a large proportion of the land area. Where
nitrogen fertilizers have contributed to the nitrate problem it is mainly
because they have built up the amount of organic matter that microbes can
break down to produce nitrate. In another context this would be described
as 'increasing the fertility of the soil' and would be seen as a good thing.
Nitrogen fertilizer is not the only factor contributing to increased concentrations of nitrate in natural waters. During and after the war substantial
areas of old grassland were ploughed to help increase food supplies.
Ploughing old grassland stimulates its microbial population to break down
the accumulated organic matter and release nitrate. Huge amounts of
nitrate are produced and these have probably contributed to the nitrate
problem at least as much as, and possibly more than, the increase in the use
of nitrogen fertilizer (Chapter 8). The wartime ploughing of old grassland
was part of the UK government's 'dig for victory' campaign and later
ploughing was part of the same government's policy. There were excellent
reasons for this policy, but it almost certainly contributed to our present
nitrate problem.

Chapter Eight
Sources of Nitrate Leakage: Grassland and Organic Farming

Grassland farming differs from arable farming essentially in that cereals are annual plants grown for seed, whereas grass is a perennial crop grown for its vegetation which is cut or grazed. Because grass absorbs nitrogen whenever mineralization is occurring, grassland is potentially a less leaky system than arable farming (Fig. 8.1). Winter cereals remove most nitrogen from the soil in spring and early summer. Mineralization of organic matter (see Chapter 4) is active all year round but particularly in spring and autumn. Only a grass crop grows strongly enough in the autumn and early spring months to intercept this nitrogen fully. Thus up to 400 kg fertilizer N per ha^{-1} may be applied to mown grass before substantial leakage of nitrogen occurs, which is twice the amount for cereals (Fig. 8.2). The other (unseen) difference between cereals and grass lies under the surface of the soil in how plants use the nitrogen that farmers give them. Figure 8.3 shows pie diagrams of the different fates of nitrogen in soil under grass and cereals. Both crops take up about the same amount of nitrogen into above-ground vegetation, but grasses direct a great deal more of the nitrogen into soil organic matter, mainly via the dense mat of roots that they build up. Cereals being annual crops, die before harvest and do not grow as much root material as grass and so do not contribute to the build-up of soil organic nitrogen to the same extent. Grass that is grown for a single year only behaves like an annual crop; indeed the carbon rich material in grass roots may immobilize nitrogen when ploughed in after 1 or 2 year's growth. Old pasture growing in association with clover that fixes nitrogen from the atmosphere, however, builds up soil organic nitrogen reserves enormously and it is this nitrogen which is released when old grassland is ploughed. Before we consider how nitrate leaches from grassland and how much does so it is useful to list the systems under which grass is managed and the ways these systems differ from arable cropping.

Of the 18 Mha of land in agricultural use in the UK in 1990 over two-thirds was grass, but almost half of this, 6 Mha, was rough pasture. Much was hillslope used for grazing sheep, or possibly deer in Scotland. There are

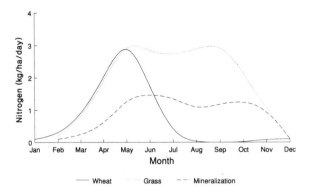

Fig. 8.1. Idealized diagram of the nitrogen taken up by winter wheat or grass throughout the year compared with nitrogen produced by mineralization from soil organic matter.

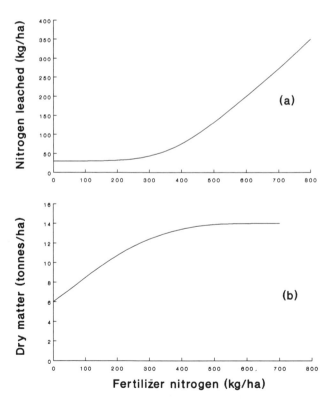

Fig. 8.2. (a) Surplus nitrogen curve for grass and (b) response of grass to applied nitrogen (adapted from Prins *et al.*, 1988).

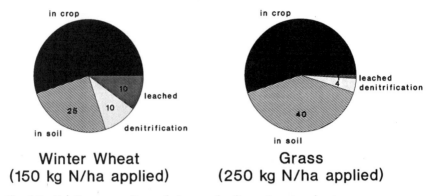

Winter Wheat
(150 kg N/ha applied)

Grass
(250 kg N/ha applied)

Fig. 8.3. Relative proportions of nitrogen fertilizer given to wheat or grass: recovered by the crop, stored in soil, denitrified or leached.

several reasons why this land has contributed little to the rise in nitrate concentrations in drinking water. First, it is relatively unmanaged, neither ploughed nor fertilized (except to a marginal extent); it also supports few grazing animals and is undisturbed except that it may sometimes be fired to discourage the growth of heather, bushes and other inedible species. Burning may increase loss of mineralized nitrogen slightly by killing and temporarily removing vegetation and by increasing erosion. Second, these rough pastures are almost always in upland counties where rainfall is high enough to dilute what little nitrate is leached to less than the European Community (EC) concentration limit. For these reasons we will not discuss rough grazing here except to say that there is some concern about increased levels of nitrate in water from these sources, because if drinking water is to meet EC standards, water containing too much nitrate may need to be blended with water containing much less. Thus it is in the interests of water companies to monitor and maintain the standard of water likely to be used in blending.

The remaining 7 Mha of agricultural grassland in the UK are divided in two again for the purposes of agricultural statistics. Just over 5 Mha have been in grass for more than 5 years and are counted as 'permanent grass'. Grass that has been growing for less than 5 years is temporary; this distinction is important. In any year it is likely that much of the temporary grass will come under the plough and revert to arable land. The permanent pasture tends to remain so.

Ploughing grass

Ploughing temporary grass releases mineral nitrogen from any organic matter that has accumulated, but the amounts released will only be 100–

200 kg ha^{-1} according to the length of the ley (Darby *et al.*, 1988). Ploughing permanent grass can mineralize vast amounts of nitrogen, about 4 t ha^{-1} over 20 years in the case of some very old grassland at Rothamsted (e.g. Whitmore *et al.*, 1991). The equilibrium in soil between nitrogen released by dead material and that taken up by growing grass is disturbed and the thick root vegetation decomposes fast because it is more exposed to air on ploughing. Much nitrate then leaches when it rains and the older the grass (within reason) the more nitrogen there is to leach out. Figure 8.4 shows how the extent of arable land has increased but grassland declined above two bore-hole catchments in Lincolnshire and Cambridgeshire during the last 50 years (Laurence Gould Consultants Ltd., 1985). These two counties both receive less than the average UK rainfall and abstract a large proportion of their drinking water from underground bore-holes; they have also undergone the intensification of arable agriculture typical of the east of the UK during the last 50 years. In 1982 permanent pasture had declined to about one-third of the amount in 1942, with the greatest decline occurring in the 1960s. These catchments are not unrepresentative of the UK as a whole. Knowing the average annual amounts of rainfall and evaporation (e.g. MAFF, 1971) and the decline in the area of permanent pasture in each county, and assuming that the net amount of nitrogen lost on ploughing is about 4 t ha^{-1} during 20–25 years, Whitmore *et al.* (1992) estimated the contribution that ploughing grass has had on nitrate leaching throughout England and Wales. Their calculations, which were used to derive Figure 8.5, assume that all the nitrogen released from soil organic

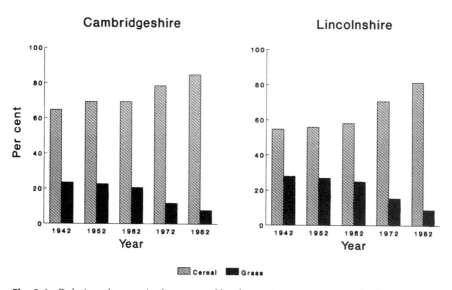

Fig. 8.4. Relative changes in the areas of land growing grass or cereals above two water catchment areas in eastern England.

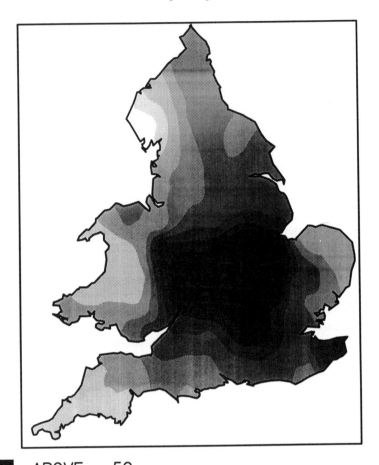

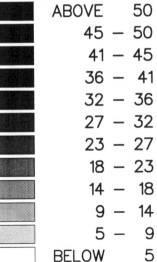

ABOVE   50
45 — 50
41 — 45
36 — 41
32 — 36
27 — 32
23 — 27
18 — 23
14 — 18
9 — 14
5 — 9
BELOW   5

Fig. 8.5. The potential contribution of ploughing *permanent* grass to nitrate leaching in England and Wales in 1945. Nitrate concentrations (mg l^{-1}) in water draining from all land assuming that the only source of nitrate was ploughed grass.

matter under the tiny area of grass ploughed each year in each county of England and Wales found its way out of the system as nitrate in the water draining from the whole county. In fact some nitrogen will have been absorbed by the crop each year and some lost through denitrification. Figure 8.5, therefore, shows the immense *potential* effect that ploughing grass might have had on the concentration of nitrate in water draining from all land in 1945. There must, however, have been other sources of nitrogen for crops during this time because the area coming out of grass was always small, with the greatest proportion ploughed in 1941 when 16% of Northamptonshire came under the plough, so leaching may have been even greater. As we have seen, fertilizers applied to arable crops are not completely consumed and they may help to maintain the soil organic matter. Some data published by Jenkinson (1988) shows that fallow soils lose more organic matter than the net loss of nitrogen from organic matter calculated by Whitmore *et al.* (1992) in deriving Figure 8.5; if the gross loss is indeed greater, more nitrate may have been leached than implied by this figure. It is far from irrelevant to show the contribution of grass to nitrate leaching in 1945, because as we shall see in Chapter 9, some aquifers retard leaching to the extent that rain falling in the 1940s is only now being abstracted as drinking water. The scale of the problem is immense and much of the increase in nitrate concentrations in aquifers under arable East Anglia and the Midlands can be attributed to the ploughing of grass.

We cannot really complain at the ploughing of grassland. The land was brought into cereal production in the UK to feed the growing population in the 1930s and to meet popular government policies designed to avoid the rationing of food that accompanied the Second World War and its aftermath. Now that this peak of ploughing has passed (Fig. 8.4) we can expect nitrate levels in drinking water to fall slightly or at least maintain their present levels. This assumes that nitrogen fertilizer levels will remain the same. Figure 8.6 shows that these were increasing inexorably into the 1980s partly to take advantage of the high yielding modern crop varieties, partly to take advantage of EC subsidies but perhaps also to compensate for the gradual decline in fertility of soil as the years after the ploughing of the grass became decades. As we saw in Chapter 1 the amount of fertilizer applied to winter wheat in the UK is now showing signs of stabilizing or even decreasing and if so, nitrate concentrations in drinking water should eventually decrease.

Grazed and ungrazed grassland

Whether temporary or permanent, grassland may either be cut and used for hay or silage or grazed directly by animals. How the grass is used in farming has a bearing on the way nitrate leaks.

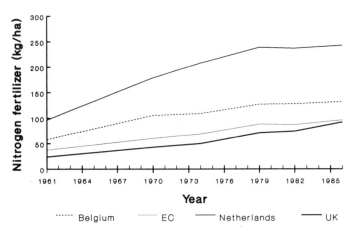

Fig. 8.6. Increase in the use of nitrogenous fertilizers within the UK and the low countries compared with the average for the EC (taken from Prins *et al.,* 1988).

Ungrazed systems

We have seen that cut grass can generally make use of up to 400 kg ha^{-1} of nitrogen, before a substantial increase in the leaching of nitrate occurs. This is because the grass is given nitrogen only as required in spring and after each cut, and because the vegetation is removed from the system. As Figure 8.1 shows, the grass is growing at a time when it can make good use of the mineral nitrogen produced from organic matter in soil. Some grasses may release substances into the soil that block the nitrification of ammonium-nitrogen (see Chapter 4). Ammonium-nitrogen leaches much less readily than nitrate in normal arable soils (Chapter 3) so this too may decrease nitrate leaching. The evidence that these natural nitrification inhibitors are common, however, is at best equivocal. It is also questionable whether grass managed like this reaches an equilibrium, and if so how long it takes. We do not know if virgin prairie in association with nitrogen fixers such as clover ever reached a balance point where the nitrogen entering the system was equivalent to the nitrogen leaving the system and both exceeded the 11.3 mg N l^{-1} EC limit for nitrate in drinking water. Power and Schepers (1989) who reviewed the scientific literature on nitrate contamination of groundwater in North America recounted observations from early explorers who found muddy rivers and polluted water holes in the 19th century in the Midwest. They also reported extremely large amounts of ammonium or nitrate-nitrogen below native prairies in the early part of this century. It may be that only our awareness of the nitrate problem is new. We do know, however, that the problem with grassland, managed or not, is that the whole grassland system becomes much more leaky if it has to support animals.

Grazed systems

The main problem with grazed systems is that they use nitrogen very inefficiently. Cut grass uses nitrogen efficiently because the nitrogen it contains is removed from the field and because it regrows after each cut and continues to grow almost all the year round. Arable crops use nitrogen slightly less efficiently because they do not grow all the year round. But the grazed system leaks most nitrate because more than 80% of all nitrogen consumed by the animal is returned to the soil in urine and dung. These returns are extremely concentrated; cows for example urinate abut 2 l at a time over an area of roughly 0.4 m². This can represent an application of between 400 and 1200 kg N ha⁻¹. Compare this with Figure 8.2 which shows that grass can safely exploit an *annual* application of no more than 400 kg N ha⁻¹. Worse still, the animals regularly urinate in the same part of the field often damaging the grass and it has been established that only 15–20% of any field is likely to be fertilized in this way. Returns of nitrogen in dung are about half those in urine but are spread on an even smaller area, perhaps 7–10% of the field. Here again the grass cannot make use of all the nitrogen, which may even be concentrated enough to be toxic. The frequent passage of animals trampling over the same spot further damages both the land and the grass. The stocking rate, the number of animals per hectare, is needed to make these calculations relevant. Although the productivity per animal in terms of meat, milk or wool, declines almost linearly as more animals are kept in a field, productivity to the hectare increases with added nitrogen fertilizer to a maximum of about six to ten animals to the hectare (e.g. van Burg *et al.*, 1981). On average, so-called itensive cattle farming supports about three cattle (or equivalent animals in terms of bodyweight and consumption) to the hectare. Grazed grass receiving up to 400 kg N ha⁻¹ can lose about 300 kg N ha⁻¹, much of it through leaching. Through-drainage of water must exceed 2600 mm a year (100 inches) to bring this loss down within the EC limit of 11.3 mg l⁻¹. Some of the loss is probably as ammonia gas (see Chapter 4), but the ammonia may only be carried to the neighbouring field or farm before it is washed out in rainfall or deposited in dust as we shall see in Chapter 9. Farmers are careful, incidentally, not to allow their animals to graze grass for 10 days or so after it has had fertilizer applied. This is because the pH of the rumen is much higher than in the human adult stomach and so, like human infants, ruminants are susceptible to methaemoglobinaemia.

Grassland that is grazed less intensively leaches less nitrate. In practice, however, many farmers actually 'graze' their animals more intensively than is implied above. Table 8.1 shows a balance made of the inputs and outputs of nitrogen to animal farming over a 12 month period within the UK. There are some objections to this form of tabulation: that it is difficult to calculate the numbers presented here from books of agricultural statistics

Table 8.1. Balance of inputs and outputs in UK agriculture.

Inputs	kt N	Outputs	kt N
Grazing	723	Meat, carcasses mortality, etc.	139
Feeds	593	Products: milk, eggs, wool, etc.	105
		Balance	1072
Total	1316		1316

Data taken from Royal Society (1983).
kt = kilotonnes.

compared annually, straddling as they do over 2 calendar years; and that the land may not be in balance at all – some land may be gaining nitrogen and some losing it. Overall, however, UK farmers have been feeding their animals almost as much nitrogen in grain and oil supplements as they have in grass. Assuming that animal production in the UK is approximately in equilibrium a nitrogen 'accountant' drawing up the balance sheet must show a loss of almost 1.1 million t N year^{-1}. Since only about a quarter of a million tonnes ends up in useful, saleable products the system is obviously very wasteful indeed. Although some of this nitrogen is lost to the atmosphere a great deal must be leached as nitrate.

The problem is certainly not confined to the UK, as is shown in Figure 8.6 which plots the use of fertilizer in the EC. Belgium and the Netherlands use a great deal more nitrogen fertilizer per hectare than the rest of the EC and they certainly use a great deal more animal feed. Indeed comparing the nitrogen imported through Dutch borders with that exported in food shows that the Netherlands ought to be accumulating nitrogen at an alarming rate.

The form of nitrogen entering the system makes no difference; be it feed, manure or fertilizer; each is used inefficiently. Pasture containing a mixture of grass and clover may be less leaky but this is because it usually introduces less nitrogen (under 150 kg N ha^{-1} year^{-1} fixed biologically from the atmosphere) and so supports fewer animals. Leakage is less, simply because an 80% loss of 150 kg N is 120 kg N while 80% of 400 kg is 320 kg. Clover growing on its own and fixing nitrogen in a grazed field has been found to leak nitrate at rates similar to grazed grassland given amounts of fertilizer nitrogen similar to the amount fixed by the clover. It is the flux of nitrogen through the animal that determines the loss. It is possible to remove animals from the fields altogether and arrange for them to be fed with mown grass or concentrates in paddocks or stalls. The problem then becomes one of disposing of the manures and slurries produced when animals are confined in small areas.

Leakage of nitrogen from manures, slurries and sewage

Manures and slurries are an excellent source of nitrogen that should not be wasted; the main problem in using them is often finding out how much nitrogen they contain. Applying them can also bring practical problems.

Farmers have traditionally applied these organic fertilizers with a view to their long term benefits. Manure helps to stabilize soil aggregates and prevent erosion; it also improves the structure of soil promoting good tilth, good moisture retention in droughty areas and, paradoxically, often good drainage in wet areas. It also supplies the soil with many nutrients, nitrogen in particular. It does this not only by stimulating the action of the soil micro-organisms (Chapter 4) that control the rate of break down of organic matter in soil but also because it contains substantial quantities of mineral or readily mineralizable nitrogen.

Different animals produce different manures because they have different physiologies, anatomies and diets. Table 8.2 shows the numbers of cattle, sheep, pigs and poultry on farms in the UK in 1989. It also shows the typical nitrogen content of manure from these animals. These are the total amounts of nitrogen contained in the manure. Only a few per cent of this nitrogen is likely to be in the mineral form and immediately available to plants, but a large proportion of the remainder will decompose fairly

Table 8.2. Numbers of animals in the UK and their wastes.

Animal	Numbers (10^6)	Dung and urea (10^{12}g)	Water (%)	Nitrogen Value[†] (%)	Nitrogen Value[†] (£10^6)	Output (g N day^{-1} animal^{-1})
Cattle	12	103	80	1.0	309	235
Sheep	43	42	65	1.7	214	19
Pigs	7.7	12.5	75	1.6	60	27
Poultry	131*	4	78	4.8	192	1.2

*1988 figures, others 1989.
[†]Assuming all g N day^{-1} animal^{-1} available and price of £0.3 kg^{-1}.
Data taken from Smith and Peterson (1982) and Smith and Unwin (1983).

readily in the soil. This table gives us much of the key information needed to analyse the advantages and disadvantages of organic manures. But first consider their financial value. Almost all pig and poultry manure in the industrialized countries of northern Europe is voided in confined areas and so must be disposed of, but its nitrogen content in the UK alone could be worth over £400 million to the farming community if it could directly replace nitrogen fertilizer.

The most noticeable thing about manure (after its odour), is its bulk. Most is water and only a very small proportion is nitrogen, so it is uneconomical to transport manure very far. Recent intensification has meant that, in the UK at least, arable farming is concentrated in the east of the country, where dry summers help to ripen grain for harvest, while animal farming has concentrated in the west, where wet summers and mild winters make the growing season longer. It is obviously uneconomical to transport several million tonnes of manure from west to east. Ideally manure needs to be disposed of as soon as it is produced; its bulk makes it costly to store. Cattle graze outside during the summer months but are kept in stalls during autumn and winter, so that the manure accumulates at this time of year. However, everything we have said so far should make it abundantly clear that the winter months are the worst ones to apply any form of nitrogen to soils. Because of this consideration and the cost of storage, transport or disposal, organic manure is often thought of as a waste material, even as a pollutant, which is odd for a material potentially worth many millions of pounds to the farming community.

Part of the problem in spreading manure is uncertainty about its composition. Its water content is rather variable, and gauging the rate at which its organic nitrogen becomes available to plants is not easy. Some American scientists suggest a kind of law of diminishing returns in a table of series (Smith and Peterson, 1982). The series (75, 15, 10, 5) for cattle manure tells us in round terms that 75% of the nitrogen in the manure will become available to plants in the first year. In the second year 15% of the *remainder* becomes available followed by 10% of the new remainder in the third year, and 5% in the fourth and subsequent years. The series for poultry droppings is (90, 10, 5). The large percentage in the first year reflects the substantial quantites of mineral nitrogen and urea present in fresh manure. Although the amounts of nitrogen becoming available 4 or 5 years after application are small, the manure has a persistent effect for many years. If farmers spread manure annually (or even more frequently) they may build up simultaneously not only the fertility of their soil, but also the risk of serious nitrate losses in draining water. An extreme example of this can be found in some parts of Europe, particularly in Germany, where since the 12th or 13th centuries farmers have been diligently improving poor heathland with manure from their animals. The result has been a layer of artificial soil from the manure, often up to a metre deep on top of freely draining sandy subsoil. In autumn these soils may contain more than 500 kg N ha^{-1} of nitrate that is in danger of leaching. With the average rainfall for central Europe this could result in a concentration of 150–200 mg N l^{-1} water leaving the soil (EC limit 11.3 mg N l^{-1}).

Another problem with manure is the losses to the atmosphere, not just the malodorous smell of the countryside but also, on dry sandy soils in particular, substantial gaseous losses of ammonia. Good practice is therefore to plough manure into the soil to reduce this loss but doing so ensures that more is in the soil at the mercy of soil micro-organisms. Greater local leaching losses of nitrate in the autumn and winter can be expected where manure is ploughed into warm soils.

Ploughing-in or spreading manure in spring rather than autumn or winter is not always a solution, because apart from the problems of storing manure over winter, there is the risk that the soil may be compacted if heavy traffic passes over it before it has dried out enough to support farm vehicles. Happily, compaction is usually less of a problem with the sandy soils expected to benefit most from applications of manure and likely to leach nitrate most easily over winter.

The composition of manures given in Table 8.2 is by no means hard and fast. One of the problems in compiling a book of this kind is choosing which of many differing sets of data to use. The values given can probably be relied upon, and a farmer using data from a table such as 8.2 is unlikely to be more than 50% out in estimating the percentage of nitrogen in the manure he applies. If he gauges the water content correctly he will probably do better still. Unfortunately, much manure is not supplied fresh from the animal but as a slurry, usually produced by hosing or washing the manure out from stalls and then siphoning it into a storage tank. Although this avoids much unpleasantness it makes it virtually impossible to assess the nutrient content of the slurry. Worse still, the nitrogen is often spread very unevenly because the solids settle to the bottom of the hopper or tank and the slurry gradually becomes thicker, and more concentrated in nitrogen as it is spread.

Arable farming only leaks less nitrate than grazed grassland because the animal (ourselves) is housed off the land, but we still have to dispose of our manure (or sewage). Human sewage was spread onto the land until quite recently in some parts of Europe. At the moment sewage from coastal areas of the UK is disposed of at sea from long outfalls designed to prevent the material washing back on to beaches. The British Government, however, is pledged to stop disposing of sewage at sea. Raw sewage in other areas is digested and processed in order to remove pathogenic bacteria and certain parasites which require an animal as the primary host before infecting humans. The risk of dysentery, cholera and so on is far greater if the organisms responsible can find their way into natural waters. Digested sewage does not present the health hazard of raw sewage and is disposed of in rivers or on farmland, but the amount of nitrogen it contains can vary by several orders of magnitude (Table 8.3) and is almost impossible to estimate. It may seem obvious to measure the nitrogen concentration and apply digested sewage accordingly, but this costs money and farmers usually take sewage because it is a free or very cheap source of nitrogen. Another problem with the sewage is that it contains a lot of metallic and other industrial waste, much of which is toxic to plants, some important soil microbes and potentially to animals. A great deal of research has been done on the persistence of the 'heavy metals' such as lead, cadmium and nickel in soils many years after the application of sewage. The compounds formed by these metals in soil are rather insoluble and they are not easily washed out. Relatively little of them is taken up by any one crop, but they are unfortunately harmful to life in very small amounts and can accumulate in animal bone or nervous tissue over many years. The presence of these toxic heavy metals is making the disposal of sewage on land an increasingly controversial procedure.

Fears about nitrates, and now pesticides, in our food and water have led increasing numbers of people to ask if modern methods of intensive farming are safe. The alternative is a return to organic or traditional methods of farming, but this raises two questions: would such a change lessen the nitrate problem, and if so, can it do so without drastic cuts in food productivity.

Table 8.3. Nitrogen content of sewage sludges.

Form of n	Digestion	Range	Median
Total N(%)	Anaerobic	0.5–17.6	4.2
	Aerobic	0.5–7.6	4.8
	Other	0.1–10	1.8
Ammonium (mg kg^{-1})	Anaerobic	120–67 600	1600
	Aerobic	30–11 300	400
	Other	5–12 500	80
Nitrate (mg kg^{-1})	Anaerobic	2–4 900	79
	Aerobic	7–830	180

Data taken from Smith and Peterson (1982).

Organic farming

Several definitions of organic farming are used, ranging from simply 'non-intensive' to the rather strict set of conditions that produce must meet before it can qualify for approval by MAFF (UKROFS: United Kingdom Register of Organic Food Standards). Most people would say that organic farmers eschew the use of chemicals but even this is not a satisfactory definition because some pesticides and some fertilizers are allowed where absolutely necessary (e.g. Chilean nitrate and rock phosphate). Since we are concerned only with nitrates in this book the most convenient description of organic farming from our point of view may be the organic farmers' own dictum 'feed the soil not the plant', for this describes the essential philosophy of organic farming very well indeed.

Conventional farmers tend to apply mineral nitrogen to their crops in large dressings in spring. We have seen in Chapter 2 that this can lead to over use of nitrogen by plants which subsequently lodge or are prey to pests or disease. Hand in hand with these applications of mineral nitrogen goes the application of chemicals to protect crops from competition from weeds and from the multitude of pests and diseases to which they are prone. This susceptibility is in part a result of the careful breeding programmes that have selected the capacity for large yields, possibly at the expense of other characteristics such as disease resistance. The inescapable fact, however, is that we now feed a great many more people than was possible even 20 years ago. The current world population could simply not have been sustained by the agriculture of 100 or even 20 years ago.

Organic farmers, however, perceive that it is not only good that has come out of the intensification of agriculture. They point out that the large inputs of chemicals to the soil–plant system upsets a natural ecosystem which ought to be disturbed as little as possible. All farming perturbs the natural ecosystem to some degree, but in principle at least, it is possible to see that feeding the soil rather than the plant causes less perturbation. Organic farmers use manures to dress crops with nitrogen or they use old methods of supplying nitrogen in crop rotation. Growing a legume crop such as peas or beans brings nitrogen into the soil because bacteria living in association with these crops fix atmospheric nitrogen (Chapter 4). Clover has the same benefit, so a field may be put down to grass with clover in it for a year or two at a time. Table 8.4 shows a typical organic farming rotation. Not all organic arable farmers have animals to supply manure and those that do not rely heavily on crop rotations. Those that have animals are normally careful to feed them only on grass and supplements, for example grain or rape-cake, that have been grown organically.

All this may sound idyllic, or naive, or healthy, or desirable but impracticable, depending on your point of view. It is hard to assess the

Table 8.4. A typical organic farming rotation and the amount of nitrogen it introduces into the soil.

Crop	Nitrogen fixed kg ha^{-1}
Ley (legumes/grass)	300
Ley (legumes/grass)	400
Oats	–
Wheat	–
Beans	250
Barley	–

Data taken from Bezdicek *et al.* (1984).

value of organically grown produce because, some honourable exceptions apart, the scientific community has largely disdained organic farming and because it is difficult to set up reliable, comparative trials on the plots of land so beloved of experimental scientists. For example, an organic plot in the middle of a conventional field benefits from an area of quarantine around it created by the use of chemicals. Experiments need to be paired so that an otherwise identical experiment is set up elsewhere with the conventional treatments surrounded by an organic farm. There still remains, however, the problem of soil variability that the small experimental plots are intended to overcome.

Thus comparisons between conventional and organic farming methods are somewhat equivocal. It seems possible that organically-grown produce contains a higher percentage of vitamins and essential nitrogen-containing amino acids, but this may simply be because the carbohydrate in the extra yield of the conventionally farmed produce dilutes the vitamins and amino acids. Also conventionally produced grain is better for baking because it contains more non-essential amino acids (Bezdicek *et al.*, 1984).

Plants must take up mineral nitrogen whether they are grown conventionally or organically. The ready availability of nitrogen from chemical fertilizers encourages fast growth and, if the weather is favourable, large yields. Organic farmers usually produce less yield of what is perceived to be a higher quality and for which some people may quite reasonably be prepared to pay a premium. To obtain yields similar to those of conventional farmers, organic farmers must ensure that their soil supplies a similar amount of mineral nitrogen. Interestingly, where this has been done, and identical yields achieved, there is some evidence that organically grown vegetables at least, contain less nitrate (Bezdicek *et al.*, 1984). However, scientists in Britain found exactly the reverse in lettuce attributing this partly to mineralization of organic manures long after application. These results may not conflict. Large yields of vegetable crops can still be obtained even with very large reductions from the current rates of nitrogen fertilizer (Chapter 10), so the scientists who found less nitrate in their organic crops may simply have supplied less to them. Crops such as spinach contain large amounts of nitrate in the morning which reduce during the day. Commercial growers therefore harvest spinach in the afternoon; but we do not know when the experiments were harvested. The British scientists may have been careful to ensure that both organically and conventionally farmed lettuce received identical amounts of nitrogen so that the crop could not distinguish whether the nitrogen was supplied organically or conventionally. Such experiments are unlikely to be relevant to cereals, and therefore to bread or beer, because there is virtually no nitrate in grain however it is produced. The availability of nitrate needed to reach higher yields under organic systems can only be achieved by applying more nitrogen in total than would be needed in conventional farming. This greatly increases the risks of nitrate leaching.

Animals reared on organically grown grass and grain are said to be more productive than conventionally reared animals. This may be due in part to a higher concentration of essential nutrients in the feed, but it is more likely to be a result of less intensive farming. Stocking rates on organic farms may be only one-tenth of those on intensive ones and we saw earlier that productivity per animal increases when the number of animals to the hectare goes down. Nitrate leaching is likely to be much less from such farms because, although the organically farmed cow voids its waste in just as small a proportion of the field as its conventional cousin, there are far fewer cows in the field, which means that the leached nitrate is diluted more as the water carrying it mixes with water draining from other parts of the field or farm.

Arable organic farms may lose less nitrate by leaching than conventional ones but this is probably only where they are less productive. The protein in the grain needs nitrogen from the soil, and to allow nitrogen to accumulate, some of the land will need to grow grass or a grass clover ley for a year or two and so be out of production. This land leaks little nitrate so the water draining from it can dilute the nitrate carried from the fields cropped annually. Overall, however, fewer fields produce edible or marketable products. There is a danger too that when these leys are eventually ploughed they may release more nitrate in the autumn than crops can use in any one year. The aim is to use fields in the ley phase to counterbalance those growing arable crops, but careless organic farmers could leach more nitrate than prudent conventional ones. However, organic farmers have been quick to appreciate the value of catch-crops

(Chapter 10) grown over winter for no purpose other than to take up nitrate that leys or other land release when ploughed. The catch-crops themselves have to be killed off by cold temperatures or ploughed-in the following spring.

Organic manures leach nitrate too. In some lysimeter trials (Table 8.5) large quantities of nitrate leaked during winter from fallow soils to which composted farm yard manure had been added in autumn. Even more nitrate leached, however, from a similar lysimeter that had received ammonium nitrate at the same time. Fortunately, most farmers recognize now that it is wasteful to apply nitrogen in autumn and that it is dangerous to leave land fallow. There are procedures, such as fallowing to control weeds, that were common in traditional methods of farming that the organic movement needs to review in order to avoid contributing to the nitrate problem.

If we wish to compare conventional and organic systems we need to ask two questions. For each unit of production (tonne of grain, litre of milk, etc.):

1. How much extra nitrogen has to be fixed from the atmosphere, by the chemical industry or by legumes, in each system?
2. How much nitrate is leached from each system?

We cannot answer these questions at present, and until we can it is questionable whether anyone can say which system is kinder to the environment.

Organic farmers, like conventional farmers, face the basic problem of achieving yields from their crops that are large enough to provide them with an adequate living, but they have the advantage of the premium paid for organic produce. This means that they can accept the loss of yield that

Table 8.5. Leaching of nitrate from organic manure.

Month	Nitrate leached mg kg^{-1} manure
November	1186
December	469
January	244
February	82
March	67
April	46

Data taken from Bezdicek *et al.* (1984).

may result from not using nitrogen fertilizer. Overall, nitrate leaching from careful organic farming is probably less than from conventional farming because it is less intensive. The difficulty is that modern methods have evolved to feed our expanding world population and it is sobering to consider, with the rate of increase of the global population rising steadily, that it was only in the very late 1980s that the UK became self-sufficient in baking-quality grain (MAFF, 1990); this despite all the bad press about grain mountains. We could not maintain this using traditional methods without using much more organic nitrogen or without bringing more land into production. Either solution will lead to more, not less, nitrate leaching. Another point to remember is that traditional farming methods depended on the large supply of cheap labour that used to be available in rural areas. Much of this labour left the land for higher wages in the factories in the 1950s and 1960s (Fig. 1.3) and many traditional skills left with it. The weekenders who took over agricultural cottages may approve of organic farming in theory but may not be keen to participate in the harsh reality that is left when traditional agriculture is stripped of the rosy glow of hindsight.

Chapter Nine
Some Complications

In this book we have asked so far: where the nitrate in drinking water comes from, what harm it does, why nitrate is used in agriculture, and in this and the last chapter we discuss how it may be prevented from entering, or removed from, natural waters. There are, unfortunately, no simple answers and there are also some complications which we need to consider, together with some larger issues both within and without agriculture that have a bearing on the nitrate problem.

One important and perhaps surprising piece of information is that all land in the UK, whether in agricultural production or not, receives up to 40 kg ha^{-1} of nitrogen each year from the atmosphere, in rain, aerosols and dust. This amount is about one-fifth of the average annual application of nitrogen fertilizer to arable crops. It is all the more startling when you realize that, if dissolved as nitrate in the average amount of water draining from arable land in East Anglia it would give a concentration exceeding the European Community (EC) limit of 11.3 mg l^{-1} nitrate-nitrogen. It is not that simple in practice. Nitrogen added to soil as fertilizer does not, as we have seen, just dissolve and wash out of the soil during winter. It takes part in a series of complex biological transformations and may not emerge in water abstracted from a bore-hole for many years. This nitrogen from the atmosphere behaves no differently. Most of the nitrogen deposited in spring and summer will be captured by rapidly growing crops, whereas some of what arrives in winter may be lost through leaching. Nonetheless, undisturbed land must eventually reach an equilibrium where nitrogen arriving (from any source) is equal to nitrogen leaving the land in any form. If sustained, this 40 kg N ha^{-1} arriving from the atmosphere each year would have to be matched by a similar loss of nitrogen from some other apart of the system, for example, by leaching. Losses also include denitrification and volatilization of ammonia so that much of this figure of 40 kg may be accounted for twice in the crude nitrogen balance above: it is part of both an input to the system and an output from it. Nonetheless, scientists investigating the 'classical' experiments at Rothamsted were consistently puzzled when they tried to calculate nitrogen balances for their experiments. These long term experiments consist of plots that have grown the same crop for many years and always received the same annual application of fertilizer. Some plots received no nitrogen at all for over 100 years (Fig. 9.1) yet the organic nitrogen in the soil is not declining while the plot yields have remained more or less constant. Nitrogen balances made on plots that do receive fertilizer consistently showed that the plants were receiving between 30 and 50 kg N ha^{-1} more than was being applied (e.g. Powlson *et al.*, 1986). The question arises then where does all this nitrogen come from?

A variety of nitrogen compounds arrive from the atmosphere (Fig. 9.2). About 10–12 kg N ha^{-1} come in rain and much of the rest in dust, aerosols or gases which are absorbed from the atmosphere by soil and plants. Amounts of nitrogen in rainfall have been increasing slowly at many sites during the last 100 years or so but it is only recently that scientists have appreciated how much extra nitrogen comes from the atmosphere (Dollard *et al.*, 1987). Agriculture contributes to nitrogen in the atmosphere in the form of ammonia volatilized from the soil and animal slurries, but this usually travels only short distances before it is brought down by rain or scavenged by crops when conditions are favourable; crops absorb ammonia in much the same way as they take in carbon dioxide, through their stomata. Some nitrogen oxides come from high temperature combustion in

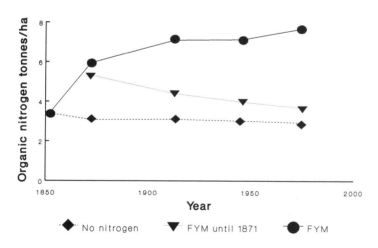

Fig. 9.1. Decline and build-up of organic nitrogen on three plots of the Hoosfield continuous barley experiment. Data taken from Jenkinson and Johnston (1977). The experiment has changed slightly in recent years such that the nil-nitrogen plot now receives some fertilizer. This affects the most recent measurement only. (FYM = farmyard manure)

Chapter Nine

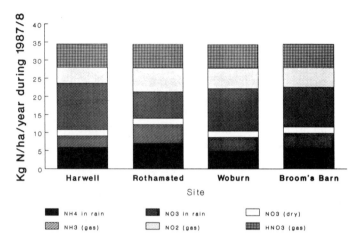

Fig. 9.2. Amounts of nitrogen from several sources deposited on agricultural land in the UK from the atmosphere at Harwell in Oxfordshire, Rothamsted in Hertfordshire, Woburn in Bedfordshire and Broom's Barn in Suffolk. Data taken from Goulding (1990).

motor vehicle engines and power stations. Some also come from natural processes in soil such as mineralization and denitrification. Figure 9.2 shows that as much nitrogen comes from dust and aerosols as from rain; this deposition supplies the missing component in the nitrogen balances on the classical experiments. It is difficult to quantify the processes that contribute to atmospheric nitrogen, assess their importance or predict the effect on them of any changes we (or the EC) impose upon land use.

As we saw in Chapter 4 nitrogen can be emitted to the atmosphere from soils and plants in three main forms: nitrogen gas, N_2, from denitrification, ammonia, NH_3, and nitrogen oxides chiefly nitrous oxide, N_2O, from both denitrification and nitrification. Fires emit mixed oxides of nitrogen but apart from stubble burning (now less common because of legislation) and forest clearing they are not a regular feature of agriculture and occur only sporadically and at point sources. Nitrous oxide, as we have seen, contributes to both global warming and the damage to the ozone layer; ammonia, which can be lost from both soil and plants, contributes to soil acidification when it falls back to land because for every molecule of ammonia oxidized to nitrate in soil two acidifying hydrogen ions are released. Clearly, any solution to the nitrate problem must not produce more of these substances.

Nitrate and nitrite

In many European countries there are now standards to limit nitrate concentrations in vegetables, particularly those given to infants. Leafy vegetables such as spinach are given large doses of nitrate right up to harvest. This is because as the leaf cells concentrate nitrate they also pull in water giving the leaf its characteristic succulence. Leaves can accumulate alarming concentrations of nitrate (up to 2% in dry matter). Also, as we saw in Chapter 1, nitrate and nitrite are added to foodstuffs deliberately because they are excellent preservatives; nitrite also gives cured meat its attractive pink colour. Our exposure to nitrite is worrying, however, not just because all the supposed health hazards of nitrate are caused by nitrite, but because it is a powerful toxin in its own right. One disturbing trend is that when a large concentration of nitrate does occur in water it is often accompanied by relatively high concentrations of nitrite. At present these are below EC safety levels but resources are probably better deployed keeping nitrite concentrations small than trying to lessen those of nitrate. Nitrite and nitrate were sampled at some 15 points in six water supply areas in the London Borough of Lambeth in 1985 (Dudley, 1986). Most of the water sampled had concentrations of nitrite below 0.05 mg l^{-1}, but there were enough above the EC limit of 0.1 mg l^{-1} (expressed as nitrite, Council of the European Communities, 1980) to cause concern and two were above 0.15 mg l^{-1}.

Much has been made of the protection afforded by vitamin C in preventing the formation of nitrite in the stomach. This implies, that vegetables should pose little threat to health, even when they contain much nitrate, because they also contain this vitamin. In some of the less developed parts of the world, however, poor soils or impoverished diets that decrease the acidity of the stomach may make this less certain. Nitrite may accumulate in the soil and be taken up by food plants. Table 9.1 shows that nitrite concentrations can also increase in vegetables stored before consumption, so we should eat fresh material where possible. Freezing or canning, however, did not seem to make the problem worse (International Atomic Energy Authority, 1984).

Historical problems

Geological nitrate

There are ways other than intensification in which arable agriculture can contribute to the increase in leaching of nitrate, through irrigation for one example. In some parts of the USA there are regions where groundwater

Table 9.1. Nitrate and nitrite in vegetables.

| | Nitrogen application rate kg ha^{-1} | | | |
	0	80	160	320
Yield* t ha^{-1}	5.1	19.4	24.4	25.2
Nitrate at harvest[†]	390	990	3260	7870
Nitrite at harvest[†]	1.6	5.5	11	9
Nitrate on storage[†‡]	330	1220	3360	5790
Nitrite on storage[†‡]	12.2	39.4	640	1075

*Fresh weight.
[†]mg kg^{-1} nitrogen in dry weight.
[‡]Transported and stored in the dark for 4 days.

moves only very slowly or not at all. Nitrate has accumulated in this water or in the rock strata that bear the water (Boyce *et al.*, 1976; Strathouse *et al.*, 1980) but the source of the nitrate is unclear. Some may have come in alluvial deposits brought down by glaciers or ancient rivers. Rocks and minerals contain much ammonium probably trapped from the primaeval atmosphere when the rocks were formed. This ammonia can be released when the rocks or soils they engender are exposed to weathering and is then nitrified rapidly.

Whatever its origin this nitrate has lain undisturbed for thousands, if not millions, of years under the semi-deserts of the USA. Irrigation has made these areas bloom, bringing agricultural prosperity to the communities living off the land, but it has also caused the nitrate to move out. Banning, cutting or even increasing the use of nitrogen fertilizer will make little difference to nitrate concentrations in the water leaving this farmland. Stopping irrigation will have a much larger effect on nitrate concentrations; it will also stop agriculture being practicable.

Denitrification

Sulphate

In some parts of Germany much drinking water is abstracted from underground aquifers overlain by light sandy soils which are used partly to grow arable crops and partly to grow conifers. The soil contains almost no clay and little carbonate. During the 1960s and 1970s much of the old pasture of the region was ploughed up to make way for arable crops. If you have deduced that this combination of circumstances should have led to a massive increase in nitrate concentrations, you are right in principle – but wrong in practice. In the aquifer at Fuhrberg, north of Hanover, and others like it a great deal of nitrate goes into the aquifer from agriculture but the maximum nitrate concentration in water pumped out from a depth of 25 m is about 1.5 mg l^{-1} well below the EC limit (Kolle *et al.*, 1985). This apparent anomaly results from a geological and microbiological quirk of nature in which the nitrate in the percolating water is reduced chemically and micro-biologically to nitrogen gas. This in itself causes a few problems because of the pressure of the gas in solution. The nitrate is reduced when iron sulphide, present in the aquifer as pyrites, is oxidized to sulphate. Sadly, this reaction is not as helpful as might appear because there is an EC limit for sulphate in drinking water (250 mg l^{-1}, 83.3 sulphate-sulphur mg l^{-1}, Council of the European Communities 1980) and the water abstracted from part of the Fuhrberg aquifer exceeds it. Iron becomes dissolved too and may manifest itself as the red colour often seen in tap water. The sulphate and the iron can be corrosive to pipe-work and industrial equipment. They also increase the hardness of water and change its nature from carbonate to sulphate during the winter; water treatment plants have difficulty coping with the massive changes in hardness that accompany the start of the leaching every autumn. Thus, although this water does not contravene the EC limit for nitrate, the nitrate from agriculture neverthe-less causes just as many problems for the water companies.

Industrial removal of nitrate

One option for water companies is to take nitrate out of drinking water with special ion-exchange plants that act rather like giant water softeners. These installations simply substitute the nitrate anion, NO_3^-, in water with another, for example bicarbonate, HCO_3^-. Eventually, however, a large quantity of nitrate-rich slurry remains that must be disposed of. It would be pleasing to think that this could be put back onto farmland to grow crops, but this is unlikely to happen because large volumes of liquid are expensive to transport. Using microbes to take the nitrate out of the water by denitrification may be a better idea, but this has its problems too. One is that the microbes need a substrate (a food source) that dissolves in water and contains carbon. Methanol is often used but must be completely consumed in the process as it is toxic to humans, thus not enough is added to remove all the nitrate; its concentration can, however, be appreciably decreased. Another problem is the gas formed. Nitrogen gas is not harmful, but the process needs to be carefully monitored to avoid the formation of appreciable amounts of nitrous oxide.

Natural denitrification

Denitrification is not restricted to soil. Water companies can usually decrease the nitrate concentration of water by allowing it to stand in large reservoirs for about 6 months. Aquatic plants remove some of the nitrate, but much is denitrified by micro-organisms in water and sediment. *Confined* aquifers, in which water runs between impervious layers and in which oxygen is limited, also support organisms that can reduce nitrate to nitrogen gas or to nitrous oxide. Current research aims to quantify this denitrification and to find out whether or not it can occur in *unconfined* aquifers into which oxygen can diffuse slowly but more readily.

Problems of water supply

The increased concentrations of nitrate in water supplies have not arisen solely from the activities of farmers. Water companies have played their part when responding to the increasing demand for water. Abstraction of water from aquifers has increased greatly during the last few years, particularly in the populous south-east England overlying chalk. Water abstraction has increased to such an extent that the water level in these aquifers is sinking and some springs have ceased to flow. This means that there is less scope for water containing too much nitrate to mix with older water containing less. Furthermore, losses of nitrate in water running off the soil surface directly to rivers and streams are diluted less as fewer springs flow. Our demand for water as consumers is therefore indirectly increasing the concentration of nitrate in the water we use.

Since little nitrate leaches out from under forest or ungrazed pasture it has been suggested that growing grass or trees on land should reduce the nitrate problem. Although growing grass may have a place, trees transpire much water and there are many parts of the world, for example East Anglia and central Europe where the annual rainfall is so low that no water at all would percolate from densely forested land. This is potentially far more of a problem for water companies.

Sewage

Looking at the nitrate problem from the water companies' point of view is instructive because it reminds us that in many countries these bodies are responsible not only for supplying water but for taking it away again as well. The population of the UK produces almost 350 000 t of nitrogen in sewage each year (The Royal Society, 1983) which is equivalent to just over 6 kg N each year from every man, woman and child. Americans seemingly produce 8 kg N per person per year (Keeney, 1989). It is possible that Americans enjoy more protein in their diet, but maybe the discrepancy reflects the difficulty in gathering such statistics. At present many water companies in the UK take advantage of the coastline to pump sewage out to sea. This will stop in the mid-1990s and will add an additional perspective to the EC limit, that of population, if you consider the likely results of spreading the nitrogen in sewage back on the land. If all the nitrogen from UK sewage became nitrate and dissolved in the average amount of water draining from soils, it would give a concentration of about 5–6 mg NO_3^- N l^{-1}, which is half the EC limit! In the USA there is much more land, but Keeney considered the nitrogen waste of a typical family to be spread onto the land they might own – half an acre. This is especially relevant where disposal takes place to a septic tank. Leaching would then be 165 kg N ha^{-1} each year; they would need about 1450 mm of through *drainage* (not rainfall) to keep this within the WHO 11.3 mg l^{-1} limit. It is probably no coincidence that the drier parts of Britain are also the most highly populated parts, and a similar analysis of nitrate losses from multi-storey blocks of flats could quickly become ridiculous. It is clear, however, that sewage spread locally on land would have a great effect on nitrate in water in densely populated areas. The rhetoric surrounding the debate on nitrate in drinking waters has ignored two simple facts. First, nitrogen cannot be destroyed or created: if we eat it, we either subsume it or void it into the environment where, unfortunately, it often ends up as nitrate. Second, it is not just farmers who put nitrate into the environment; each of us is responsible annually for more than 6 kg of organic nitrogen that is easily converted to nitrate.

Anticipating nitrate – prediction

Aquifers

The problem with predicting future nitrate concentrations in water from changes in agricultural practice is that much of the water has already fallen as rain, percolated through soil, carrying nitrate with it, and is now stored in aquifers. In some areas where soil overlies chalk, water remains in aquifers for a long time (50 years or more) while, for example, in fissured limestone or sandstone the residence time is much less, perhaps 5 years or so. This means that changing agricultural practice now will not necessarily make any difference to nitrate concentrations in bore-hole water for many years to come. Where the water now abstracted from bore-holes fell 50 years ago it is relatively easy, but expensive, to predict the concentrations of nitrate in future abstractions by boring into the aquifer rock and measuring nitrate in the water (e.g. Foster *et al.*, 1986). Hydrogeologists who did so in successive years found that most water moves downwards through chalk at a rate somewhere between 0.5 and 1.5 m a year with a small proportion moving very much faster. The proportion moving faster through a fissured limestone is much greater. This water will arrive at the point of abstraction during the next few years. Although the long time-scale makes it difficult to attribute a specific cause to the nitrate problem over aquifers, it makes historical information such as that relating to the ploughing of grass over the last 50 years all the more relevant. It also highlights the importance of research into denitrification within aquifers.

Surface waters

Our drinking water comes not only from deep bore-holes, but also from relatively shallow aquifers or rivers. In this respect, however, farming has been able to restrict bad practice and alleviate the problem. Until recently farmers applied dressings of nitrogen to winter crops in autumn. When it was shown that their crops made very little use of this nitrogen many of them stopped and in the last few years there has often been a decline in nitrate concentrations in surface waters and in aquifers that respond quickly to rainfall (e.g. Croll and Hayes, 1988). Even so, dressings of nitrogen made too early in the year are at risk, as are slurries and manures which still tend to be applied in the autumn. Even when incorporated into soil the latter are at risk of being lost: to the atmosphere where they may be washed out quite quickly onto other land, by downward leaching, or by runoff from the surface of very wet land into rivers and streams. This last probably does not represent a serious hazard to human health unless occurring in a river just above an abstraction point or unless the nitrate seeps into a drinking water well. It can, however, be quite serious for the rest of the environment. Quite apart from other ecological changes it causes, in freshwater systems surplus nitrogen leads many aquatic plants, such as reeds to take their nitrate from the water rather than the bank and thus put down feeble root systems that are not strong enough to anchor the plant when water flow increases. River banks may become eroded as the reeds are washed away loosening the soil and probably bringing more nitrate (and also phosphate) into water from the organic matter in the eroded soil. Water courses near the coast discharge this nitrate into the sea where marine algae take advantage of the extra nutrients to grow to excess. Interestingly, quick growing reeds are now proposed as a means of removing nitrate from reservoirs or rivers.

How much does the nitrate problem depend on us?

There was a story (possibly apocryphal) current about 10–15 years ago that one of the 50 states of the USA was not considering measures to contain the nitrate problem, even though the other 49 admitted to being very worried about it. This was all the more odd because the state in question was one of the most advanced and was in favour of agricultural intensification. On closer investigation the reason became apparent: while 49 states had adopted the WHO recommended limit of 11.3 mg l^{-1} as standard, the one state had adopted a limit of 22.6 mg l^{-1}. This makes the point that the nitrate problem depends appreciably on current perceptions; perceptions sometimes change. Today's problem, that of pollution by nitrate, is to a large extent the result of microbes breaking down the soil's organic matter and releasing nitrate. A little over 20 years ago the break down of organic matter itself was at the centre of concern, for a quite different reason. It was feared that too great a decline in organic matter brought about because farmers were supplying *too little* nitrogen fertilizer would result in deterioration of the soil's structure, leading to compaction in some soils and erosion by wind or water in others. Either problem would have lessened yields.

Work on the classical experiments at Rothamsted suggests that this was an unnecessarily pessimistic view. Look again at Figure 9.1 which shows some results from the Hoosfield experiment that has grown barley every year since 1852. Until recently one plot had received no nitrogen at all since the experiment began, while another has been given farmyard manure every year. A third plot received farmyard manure from 1852 until 1871 but no nitrogen (or other fertilizer) from then onwards. As we saw earlier, the plot receiving no nitrogen has not lost organic matter, nor have yields decreased on it. The grain harvested each year contains almost 20 kg N ha^{-1} each year which must come from the atmosphere; alternatively nitrogen released from soil organic matter may have been replaced by nitrogen from atmospheric deposition. The amount of organic matter in the plough depth of the soil has remained stable at about 3 t ha^{-1}.

Nitrate is a legislative as much as an environmental problem. We are charged by EC law to remove from our drinking water a substance that is not usually harmful to human health, but which can harm parts of our wider environment. We must clean up where we can, but the problem has built up over many years and will not respond quickly to control measures. We should not see this as an excuse for inaction, because each of us contributes to the problem. There are, fortunately, some measures that should help us to control the problem and these are discussed in the next chapter.

Chapter Ten
Ways of Decreasing Nitrate Leaching

The developed world is becoming increasingly concerned with environmental issues, such as the depletion of the ozone layer, climate change and chemicals in food and water. The problems we now face in the environment stem mostly from over-production or over-consumption of natural resources. The increased nitrate we now find in our water, food and environment has come about through intensification of agriculture: in ploughing up old pastures or consuming energy to synthesize nitrogen fertilizers. In some parts of the world nitrate in tap water may have come from land that had not previously received sufficient rain to cause percolation for millions of years but which is now irrigated. In each case the changing intensity of land management has brought about the increase in nitrate leaching. In this, the last chapter, we shall consider some of the general ways of decreasing nitrate in water and then ways that will be effective for a particular crop or system of farming. At the end of the chapter we shall discuss what measures the individual can take to reduce his or her exposure to nitrate and how we may each go about decreasing nitrate in the wider environment.

Agricultural controls

Almost any agricultural activity leads to some leakage of nitrate. Leakage from land under natural vegetation is almost always less than from land that is managed in order to produce something. Thus abandoning agriculture altogether would lead to a decrease in leaching and less nitrate in water. As with most simple but drastic solutions to a problem, the solution raises a series of further difficulties. First, it would bankrupt farmers. Perhaps society would be willing to defray the cost to the individual farmer, perhaps not. Second, a nation adopting this policy would soon be short of food. It might import food from a foreign country not so fussy about nitrate in drinking water, but it would simply have exported its nitrate problem.

Furthermore, it hardly makes economic sense to pay for large quantities of food from abroad that could quite easily be grown at home, while putting the work-force that could grow it onto income support from the state. The European Community (EC) was notorious with the popular press in the mid-1980s for over-production of agricultural goods. But the stores of grains were never mountainous and although traders and farmers hold healthy stocks (at the time of writing), wheat currently held represents only 10 days' supply or about one-tenth of what it was in 1985 (MAFF, 1990). Ten days' supply is certainly not excessive, and from another standpoint represents rather a meagre reserve.

Although the wholesale abandonment of agriculture may not be a viable option for solving the nitrate problem, selective abandonment of a proportion in a sensitive catchment area may well have a place. It may be possible to design a patchwork of areas leaking more or less nitrate within a catchment that enables agriculture to continue normally within some patches. To manage this it will be necessary to have reasonable estimates of the quantities of nitrate leaching from land under many different crops, and from the abandoned areas. By using mathematical models, based on sound data, it should be possible to produce a range of cropping patterns for a given catchment.

Abandonment need not be permanent. Apart from the aesthetic aspects, it could be disastrous to allow land to become derelict after centuries of careful management at a time when the world's population is increasing rapidly and fluctuations in the price of oil are making the manufacture of chemical fertilizers less viable. Most soils in arable use in the Middle Ages were chronically deficient in organic matter and nitrogen. Farmers eagerly collected such diverse materials as pigeon droppings and seaweed to compost or spread on land to replenish the soil's fertility. Crop rotations have always been a part of agricultural systems if not to control weeds and disease then because some land cannot supply large quantities of moisture or nutrients every year. There is no reason why rotations should not be used to control nitrate leaching but we must select cropping systems which neither fix too much nitrogen from the atmosphere nor require extra nitrogen from fertilizer for growth.

In many European countries, there are now recommendations or regulations stipulating the maximum area of land that may grow wheat or oilseed crops or graze animals above water catchment areas. In Denmark by law 65% of land must be cropped during winter (reducing the amount of fallow before spring-sown crops) and farms having more than 30 cattle or the equivalent in other animals must have storage for the amount of manure that could be produced in 9 months (Hansen, 1989). In sensitive areas in Germany land can receive an average of no more than 186 kg N fertilizer ha^{-1} each year but this is taken in rotation so that crops that require much nitrogen can be balanced by those requiring less. In the UK farmers tilling land overlying certain bore-holes – the so-called nitrate sensitive areas (NSAs) – have been invited to join a voluntary scheme to decrease nitrate leaching. They may join at one of two levels: the basic scheme pays between £50 and £100 ha^{-1} to farmers who agree to use nitrogen fertilizer at rates a few per cent below the optimum and maintain green cover throughout the winter. The premium scheme pays up to £380 ha^{-1} (depending on location and extent of participation) to farmers willing to convert arable land to low-intensity grassland or woodland. Indeed throughout the EC there is a system of grants to encourage farmers to set land aside from farming for a period of 5 years, the so-called *set-aside scheme*. Grass must be sown and cut twice a year but it may not be grazed. This is in sharp contrast to the policies of the UK government in the fifties which provided fertilizer *subsidies* for farmers who ploughed out grass (MAFF, 1951). As a result of these subsidies some land only marginally suitable for arable-use was unfortunately brought into production. Its very unsuitability sometimes meant that crops did not grow as well as they might and so did not make best use of soil nitrogen.

Education and advice

Much can be achieved simply through education. In part this is what we have sought to achieve in writing this book. There is little point, say, in one farmer squeezing his profit margins to the limit by cutting fertilizers and yields if the majority of the nitrate on his side of a watershed flows harmlessly away from points of abstraction of water. The neighbouring farm on the other side of the watershed may contribute much water to local drinking water supplies and so measures taken on this farm will have much more effect.

Perhaps the greatest effect can be had simply by following sensible farming practice. In temperate climates it is foolish to apply nitrate fertilizer to land covered with snow or frozen; when the snow melts it carries the nitrate across the surface of the land, away from the crops for which it was intended and into water courses. This seems obvious, but the winters are so long and the need to dispose of animal manures so great, that it has been common practice in central European countries to spread slurries on frozen land. In tropical areas farmers eagerly awaited the wet season but nitrate applied beforehand is very likely to be washed away as soon as the rain does come.

Farmer education does not have to be simply qualitative; in the USA the Environmental Protection Agency (Aller *et al.*, 1985) has developed a measure of nitrate (and pesticide) leaching risk known as the DRASTIC index.

(D) epth to water table	5
(R) echarge rate of aquifer	4
(A) quifer medium	3
(S) oil medium	2
(T) opography (slope)	1
(I) mpact of unsaturated soil	5
(C) onductivity (hydraulic)	3

This chart shows the weight given to each of the components making up the index. Thus the speed with which water moves through the unsaturated zone between soil and aquifer has the greatest effect along with the depth to the water table. The slope of the field has least weight and least effect on the index of leaching risk. A field's score on the DRASTIC scale is obtained by multiplying the weight against a field rating for each component in turn. The ratings may be obtained from two sources: the slope and soil texture are held on a database known as the National Resource Inventory and the other components of the system can be obtained from county-wide averages. The score is then the sum of all these weighted ratings and the higher the score the greater the risk. The US Department of Agriculture (Alexander, 1987) and the Soil Survey and Land Research Centre in England and Wales (Jones *et al.*, 1989) operate similar schemes based mainly on maps of soil texture. That is to say, the more clay or silt a soil contains the less likely it is to leach nitrate.

Much research is going into the development of computer models that predict nitrate leaching and the effects on this of incorporating straw or catch-crops and the turnover of organic matter in soils. Knowing the effects of these factors enables better estimates to be made of the supply of nitrogen from the soil and its use by the crop. These give a clearer picture of the overall nitrogen balance in the soil and thence a more satisfactory estimate of how much fertilizer needs to be supplied to crops to maximize yields. To minimize nitrate leaching we also need to know the 'surplus nitrogen' curve discussed in Chapter 2. Farmers should fertilize their crops to be on the horizontal part of this curve. That is to say they should apply just enough to obtain good yields, but not chase after the last few tenths of a tonne of yield that can be obtained only very inefficiently.

These are general measures. To illustrate what is being done and what can be done we will now consider the main agricultural systems one by one and discuss specific measures, both management and scientific, which we feel will be most effective in lessening nitrate leaching. It is striking that what works well for one crop system will not necessarily work well for another. It is also worth bearing in mind the scale of the problem. Most land in the world that is used for agriculture grows either a cereal, some other staple food or grass to support animals. Measures designed to reduce leaching from under cereals or grazed grass will therefore have the greatest effect on water supplies that are drawn from a large area. In other words, even a large amount of nitrate leaching from a spinach field will matter very little provided the rest of the water draining into the aquifer or river is low in nitrate. Most farmers are anxious to decrease nitrate leaching from their land if they possibly can. Naturally they are equally anxious to stay in business. They are understandably confused at the conflicting advice being proffered and upset at being held responsible for a problem largely beyond their control. We have to ask: what are the best measures any individual can undertake? How best can a farmer reduce the impact of nitrate without suffering ruinous loss of yield? In some cases there is a single, simple solution but in most there is not. A package of measures is needed, perhaps including restrictions on the amounts of risky crops in sensitive areas, a policy of sowing rather more grass than is ploughed-in, more precise advice on the nitrogen fertilizer requirements of a crop on a field by field basis and investment in scientific research into understanding and quantifying the nitrogen cycle to plan agriculture for the future.

Problem areas

Arable land

Some restriction of the use of nitrogen fertilizer in sensitive areas is wise. Intensification of farming has led to many fields growing winter wheat year after year after year. Large yields are maintained and the effects of the diseases that are passed on from one crop to the next are partially mitigated by large dressings of nitrogen fertilizer.

Although nitrogen fertilizer is cheap, nitrate leaching means nitrogen wasted and that farmers can conserve money by saving fertilizer. By sowing winter crops early rather than later they can make the best use of nitrogen mineralized from soil organic matter during autumn. Figure 10.1 makes exactly this point. The data are from experiments at Rothamsted (Widdowson *et al.*, 1987) but the same conclusion has been reached for agriculture in most temperate countries. Sowing a winter crop early allows it to grow roots sooner and deeper than a crop sown later. The catch is that a crop sown early is more prone to disease in mild autumns than one sown later and may need pesticides to achieve good yields. In the results from most of the experiments shown in Figure 10.1 the early crop was sown in mid-September, the later one in mid-October and diseases were controlled with chemicals. One month clearly makes a great deal of difference. Sometimes, however, farmers cannot sow another crop straight after harvest. They may then try to limit nitrate leaching over winter by incorporating, that is ploughing-in, straw.

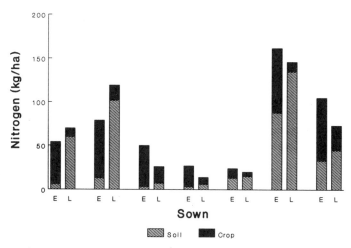

Fig. 10.1. The amounts of nitrogen taken up over winter by several winter wheat crops sown in early (E) autumn or later (L) compared with the mineral nitrogen left in the soil (Widdowson *et al.*, 1987).

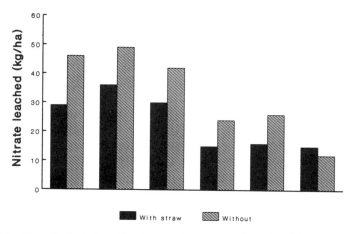

Fig. 10.2. The effect of ploughing-in cereal straw on nitrate leaching in the following winter after harvest (Jarvis *et al.*, 1989).

Figure 10.2 shows that about 50 kg ha^{-1} of nitrate-nitrogen will leach during winter following a cereal crop, but that ploughing in the straw can reduce this loss. This may seem strange at first sight because cereal straw itself contains nitrogen. However, it contains so little, compared with its carbon, that soil microbes must use indigenous soil nitrogen that would otherwise leach as well as nitrogen from stubble or roots in order to break down the carbon compounds in the straw (see Chapter 4). The results in Figure 10.2 come from a number of experiments in different years and on different soils, but they are in almost complete agreement in showing that leaching can be reduced by up to 30 or 40% by ploughing straw into the soil. But in the long term, this adds to the organic nitrogen reserves of the soil and may cause *more* mineralization and leaching (e.g. Powlson *et al.*, 1987).

Another way in which farmers can decrease nitrate leaching during winter is by sowing cover- or catch-crops to lessen the time for which the soil is bare. Catch-crops are intermediate crops that have little or no commercial value but take up nitrate effectively. Originally a catch-crop was one like turnips that farmers grew through the winter and then allowed sheep to graze in early spring; but for the most part a cover-crop is simply used to take up and store nitrogen over winter that would otherwise be leached. In this way the soil is not left bare during the critical winter months. Crops such as rape or legumes are normally harvested in mid-summer, too early to sow a winter cereal, so the farmer sows a crop of rape, white mustard, radish or rye-grass, which is ploughed in before wheat is sown in the autumn. If a spring crop is due in the rotation, the benefits of the catch-crop are even more marked because the soil is bare much longer. The catch-crops vary with the efficiency with which they can take up nitrate and with which they release it again when ploughed-in. Some crops are not hardy enough to last all winter; if these die in early frosts they may release their nitrogen early enough for it to be leached over winter. Thus much effort is being put into finding the best crops for each climate and crop rotation. As we have seen, ploughing-in cereal straw before the next crop is drilled can be beneficial because wheat straw contains little nitrogen relative to carbon. The same is not necessarily true of rape or legume residues which may contain much readily decomposable nitrogen. These should only be ploughed in when the soil is cold or when another crop is going into the soil soon afterwards to make use of the nitrogen they release. Catch-crops are particularly useful after oilseed rape and other *Brassicas*. Figure 10.3 shows just how much nitrogen some different crops can take up during winter. Most of the crops shown took up more than 50 kg N ha^{-1} and the best, white mustard took up about 100 kg N ha^{-1}. A key point, however, is whether this nitrogen is made available quickly enough to the commercial crop that follows it, or if there is a reduction in the yield of the commercial crop for any other reason. In droughty areas, for example, the catch-crop may take from the soil water needed by the commercial crop in spring. Farmers must be careful too, that the catch-crop does not provide a haven for diseases that decrease yields. Much research is currently being made to determine the best sequences of commercial and intermediate catch-crops. Figure 10.4 shows the results from a few experiments made to assess the yields of sugar beet grown after a catch-crop. It shows the amounts of nitrogen in the sugar beet and the catch-crop when the latter was ploughed under before beet was drilled. The

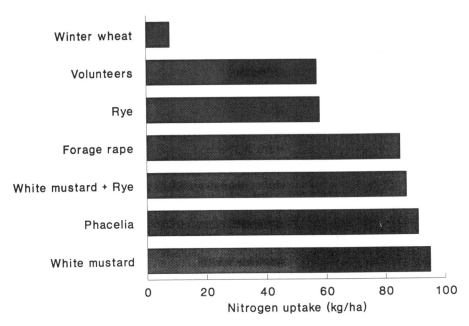

Fig. 10.3. The amounts if nitrogen taken up by some typical catch-crops over winter (Christian *et al.*, 1990).

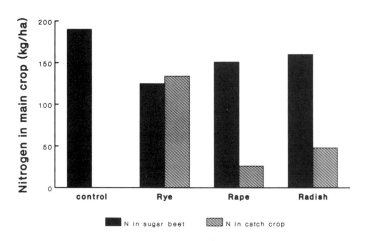

Fig. 10.4. The amounts of nitrogen remaining in soil after growing three catch-crops and the amount of nitrogen from these same crops recovered by a sugar beet crop sown afterwards (Muller *et al.*, 1989).

results show, unfortunately, that where the catch-crop removes most nitrogen from the soil, the nitrogen in the sugar beet is decreased the most compared with a control with no catch-crop. After radish, however, which removed about 50 kg N ha^{-1} the nitrogen uptake of the beet was barely reduced. Other experiments also suggest that of the nitrogen caught by the intermediate crop only about one-half becomes available to the commercial crop. Much subtle quantitative work remains to be done here. A further problem is that denitrification is often limited in soils by the availability of carbon. Large additions of carbon in the form of catch-crop residues may increase the loss of nitrate by this pathway. This does not make economic sense and if much of the greenhouse gas, N_2O, is produced it does not make environmental sense either.

Nitrification inhibitors have been used in the USA for some years to keep ammonium-nitrogen in soil as ammonium-nitrogen during autumn at least until the soil freezes. In the USA the problem is less one of leaching but more one of trafficability in spring because the soil is wet and susceptible to damage. Farmers apply nitrogen in autumn so that it is there in spring when crops need it after the thaw. They do not wish to lose nitrogen through denitrification so they apply nitrogen as ammonium together with an inhibitor to prevent nitrate being produced. Some scientists are hoping to use these same inhibitors in temperate climates to prevent the nitrification of indigenous soil ammonium formed during the mineralization of soil organic matter.

One interesting scheme is shown in Figure 10.5. Farmers do not normally get their best yields from headlands (the field perimeters).

Fig. 10.5. An idealized field plan with grass surround showing how the cereal-growing area may be reduced by approximately 10% without as much reduction in yield.

Tractors and sprayers turn or criss-cross over these field margins double-drilling or damaging plants with wheelings. By putting say a 15 m surround down to grass in a 10 ha field a farmer would lose 10% of the field but somewhat less than 10% of the yield. With a little care when applying nitrogen, however, the fertilizer application could be decreased by almost the full 10% and cut nitrate leaching by rather more as the grass surround built up soil organic matter. There is even a suggestion that such grass guard strips could harbour natural predators for crop pests and that these could balance any yield depression from weeds by reducing the farms' expenditure on pesticides. All in all farmers could see little loss of profit, cut nitrate leaching by at least 10% and with approval from the government even qualify for the set-aside scheme! This idea is not new. William Cobbett commented on such 'hedge greens' in his book *Rural Rides* published in 1830.

Grassland

We saw earlier that much of the nitrate leached from grazed grassland results from an uneven distribution of nitrogen excreted by grazing animals (Chapter 8). The nitrogen in slurries and manures can probably be spread more evenly and in more reliably known amounts if it is analysed chemically, but the main scope for improvement is in the evenness of spread of nitrogen from the grazing animal. Careful or more frequent rotation of grazing could do much to spread the areas of a field where cattle congregate to 'chew the cud' and no doubt void waste nitrogen. Electric fences are portable and can divide a field easily; more frequent rotation of grazing areas might decrease the troublesome incidence of 'poaching' when cattle trample over the same part of a field time and again causing compaction of the soil. The measure that would almost certainly have the largest overall effect, would be to farm less intensively. This implies regulations and quotas, and these tend to alter economic margins suddenly and on a nation-wide scale with the result that some regions intensify while others abandon farming altogether. This has already happened to some extent with the introduction of milk quotas in the EC. The result of this well-meaning policy was that the bigger farms bought the quotas of the smaller ones and intensified on a local scale even though milk production in the UK (and throughout the EC) decreased overall. This could well mean that the nitrate problem in the intensively farmed counties became worse. Voluntary codes or legislation need to be thought out carefully to ensure that what we do to reduce the nitrate problem does not make that problem or any other worse or simply move it from one place to another.

Grass and clover systems are generally less intensive than those involving nitrogen fertilizer but because nitrogen is free (after the cost of the clover seed) they can be almost as profitable as systems using fertilizers and heavy mechanization to spread them. They also leach less nitrate if managed carefully (Fig. 10.6). Halving the 420 kg ha^{-1} of nitrogen applied to grassland grazed by cattle halved the nitrogen returned to the soil in excreta by the grazing animals. Replacing the fertilizer nitrogen completely with clover gave returns of nitrogen broadly similar to those from the half rate of fertilizer nitrogen; the returns per animal were slightly smaller but those per hectare slightly larger. Part of the reason for the differences was that the larger nitrogen rate supported more animals to the hectare, but worse from an environmental point of view, was that the animals it supported voided more *concentrated* nitrogen in urine and dung than on grass given less nitrogen. The total nitrogen voided per hectare per week was three times larger on the plot receiving 420 kg N ha^{-1} than the one grown with clover.

We must not confuse pollution from farm accidents with that from bad farming practice, but accidents do lead to contamination of water sources by nitrate. Nitrate can, for example, leak into natural waters from badly stored slurry, and poor silage towers can leak an effluent that contains easily mineralized nitrogen. Legislation is already in place to control these events but inspectors are few. The most effective means of preventing these accidents is in the design and availability of cheap but effective storage tanks and reliable means of transporting nitrogenous effluents.

Some current research programmes are investigating the diets of farm

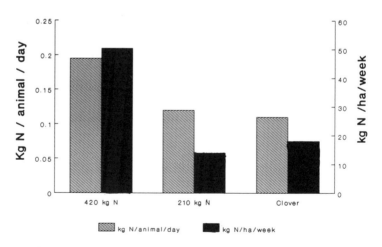

Fig. 10.6. The amounts of nitrogen returned in excreta by cattle grazing rye grass that was supplied with nitrogen in three different ways (Jarvis *et al.*, 1989). Left-hand scale shows the return of nitrogen to the soil per animal per day; the right-hand scale (kg N ha^{-1}) accentuates the differences between experiments because the high nitrogen treatment (420 kg N ha^{-1}) supports more animals to the hectare.

animals. Many of them may receive too much protein, and just as straw helps immobilize indigenous soil nitrogen, animal supplements that are rich in carbohydrates for energy and maintenance but poor in protein may help to decrease the concentration of nitrogen in the excreta and increase the efficiency with which nitrogen is used. They will, however, decrease the cost-effectiveness of farming methods that produce protein in the form of meat or milk. Scientists are examining the ways in which the animal uses nitrogen and proteins with a view to maximizing the incorporation of the nitrogen from the feed in protein and minimizing the return to the soil. Research on this issue is critical to avoid increasing the production of methane (a greenhouse gas) which could occur if the animal were to use *carbon* less efficiently in the absence of nitrogen.

Other research has quantified leaching of nitrogen from cattle urine and shown that little nitrogen leaches from urine spread on grass during summer. Through September, October and November, however, the amount that leaches increases because there is little growth of grass to use it (Fig. 10.7). In a series of experiments made in Ireland (Sherwood and Fanning, 1989) cattle urine was collected from the animal, analysed for urea and ammonium and then applied to different plots in known amounts throughout the summer and autumn. Compared with a control receiving no urine, any nitrogen in urine applied before September left virtually no trace in the soil mineral nitrogen by December (Fig. 10.7), whereas that in urine

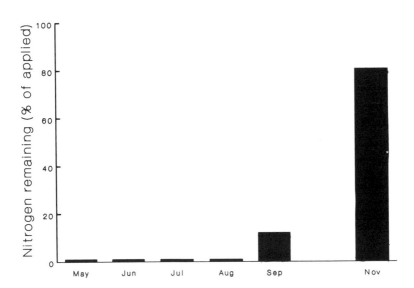

Fig. 10.7. The amounts of nitrogen remaining in soil in December (and therefore at risk of leaching) from cattle urine spread on land on different dates in the 6 months immediately beforehand (Sherwood and Fanning, 1989).

applied in September or November was clearly detectable. Over 80% of nitrogen from urine applied in November 1985 was still present in the soil 1 month later. The amounts of nitrogen in grass growing on these plots showed that the nitrogen in urine applied early in the year was taken up into herbage, but that applied later was not because the grass grew so little. A repeat experiment in 1986 confirmed these results. In areas with a known risk of nitrate leaching some cattle could be taken off the land earlier than at present, but only provided there is an effective means of catching their wastes and storing them until they can be safely spread on land in the spring.

Horticulture and forestry

Figure 10.8 compares the ways in which various, mostly vegetable, crops affect the distribution of nitrogen at harvest. Deep rooting crops, such as cabbage and sprouts, virtually empty the soil of mineral nitrogen at harvest. These are both members of the *Brassica* family that also includes oilseed rape which can be expected to deplete the soil of mineral nitrogen similarly. Other crops such as peas, that introduce nitrogen by fixation, have poorer roots and do not take advantage of mineralized nitrogen deep in the soil. Figure 10.9 compares the average rooting depth of a few arable and horticultural crops. Spinach has a poor root system and requires large quantities of nitrogen to produce the leafy green tissue for which it is sold. The leaf takes most of this nitrogen away from the soil when spinach is harvested leaving very little to be mineralized afterwards from crop residues such as roots or stalks. Figure 10.8 also shows the way in which

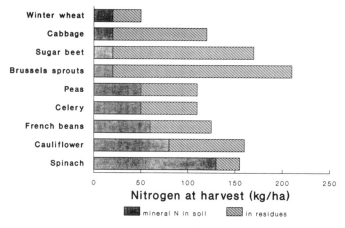

Fig. 10.8. Typical amounts and distribution of nitrogen in soil and crop at the harvest of a number of crops (Wehrmann and Scharpf, 1989).

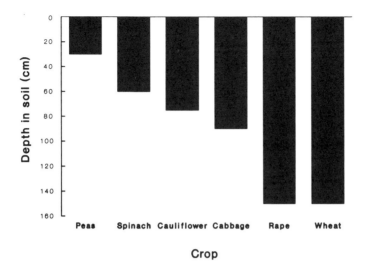

Fig. 10.9. Typical depths to which the roots of some common crops grow (taken in part from Wehrman and Scharpf, 1989).

these same crops affect the potential amounts of nitrogen that can be mineralized after harvest from their residues. Wheat in this study left only about 30 kg N ha^{-1} in residues (see also Chapter 7) but the *Brassicas* left large residues. Sprouts and oilseed rape can leave up to 200 kg N ha^{-1} in a readily mineralizable form. Leafy vegetables and legumes leave residues containing intermediate amounts of nitrogen in soil, but when mineral nitrogen is included, most leave over 100 kg N ha^{-1}. Cereals are easily the best of these crops to grow to limit the amount of nitrogen in soil at risk of leaching.

In the past the nitrogen in the residues of the first crop of the year has not always been included in the calculations of the amount of nitrogen that must be supplied to the second and in some climates the third. This must be done because the usual conditions for growing vegetables, irrigation and warm weather, are ideal for soil micro-organisms to produce mineral nitrogen; this, therefore, is one system of farming where the N-min method (Chapter 2) works very well indeed. Growers can have soil cores taken to measure the amount of mineral and readily mineralizable nitrogen in soil or they can use ready reckoners of the kind developed at the Institute of Horticultural Research, Wellesbourne (now Horticultural Research International).

As always the scale is important. One field of allotments in a high rainfall district will cause only a fraction of the overall nitrate problem, but a wide valley full of market gardens above a water catchment for a large town poses an altogether different problem. Most growers in temperate climates can get two or three crops in a year. It makes sense to grow first the ones leaving the largest nitrogen residues or requiring the largest concentrations of nitrogen in soil and follow these with the ones that root deepest. This common sense approach may be over-ruled by other demands of the plants such as long growing days or the absence of frost, but it is a principle worth following where possible. Vegetable residues contain much easily mineralized organic nitrogen so it is wise to incorporate them into the soil late in autumn rather than early because soil micro-organisms become less active as the soil gets colder and cannot decompose them so quickly.

Vegetables require and respond to, large doses of nitrogen; this nutrient more than any other produces a luxuriant growth of green leaves. Research in Spain has shown that vegetable yields can be maintained by applying less nitrogen overall and splitting it into several small applications (Fig. 10.10). Very large amounts of nitrogen are normally applied to some vegetables, and not just in Spain. Organic manures are sometimes applied as well as mineral fertilizers. In one or two instances more than 1000 kg N ha^{-1} was applied in manure and followed by inorganic nitrogen! The decrease in soil mineral nitrogen in the treatments with less nitrogen (Fig. 10.10) is not always 50% because many of the farmers also applied organic manures to both treatments. The total amount of fertilizer nitrogen applied (indicated

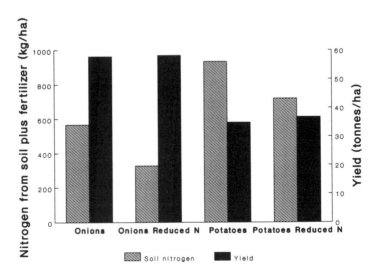

Fig. 10.10. Showing how some crops receive more than enough fertilizer nitrogen (Ramos *et al.*, 1989). In the reduced nitrogen treatment the onions or potatoes received half the quantity of inorganic fertilizer in total given to the control plots. Reduced dressings, however, were given twice as often. The yields of the crop were not reduced.

by the scale on the left-hand axis) in the reduced treatment was half that applied normally. Figure 10.10 shows quite clearly that there was no reduction in yield from applying less nitrogen and the scientists have recommended their system be adopted by farmers in Spain. This is a clear illustration of the use of the 'surplus nitrogen' curve; these vegetable growers are supplying more nitrogen to their crops than is really needed. Less nitrogen does not bring about a reduction in yield, although to be absolutely sure growers can keep nitrate concentrations in soil high by giving fertilizer more frequently. Placement of fertilizer just below the crop is another way of decreasing the amount applied without decreasing the yield.

Certain crops, like potatoes, give farmers the worst of all worlds. They need large amounts of nitrogen to grow enough leaves early in the season, but they have shallow, poorly developed root systems, ill-equipped to take it up. They are grown in ridged-up soil to make the tubers easier to lift, but this means that late dressings of fertilizer often get spread in the furrows away from the roots. Worse still the crop is almost always irrigated and usually grown on light land from which the large amounts of nitrate released from the nitrogen-rich residues potatoes leave can easily be leached.

Land down to forestry usually leaks very little nitrate until the trees are felled. Soil organic matter levels are often higher than in arable or fallow land and can release large amounts of nitrate, so felling a large area of trees can have a strong impact on the environment. The solution is to clear only small areas of trees at a time so that any nitrate produced has a chance either to be captured by other trees or to mix with low nitrate-water from the area around. Simply felling one tree in 20 is a good idea but this is a problem where stands are densely packed.

What the consumer can do

Everyone blames farmers and farming for nitrate in water. They have, however, only been doing their job – feeding the population. There was a time in the recent past when farmers made much money in so doing, by taking advantage of Government or EC subsidies. The resulting rapid intensification of farming must have contributed to the problem, but if so, the problem was caused by politicians as well as farmers. Times have changed: it is unlikely that many arable farmers are showing a profit at the time of writing and current farming practice seems likely to stabilize or even reduce nitrate leaching in the long term. It will probably not be enough to enable some water suppliers to comply with EC directives in the short term, however; and if we are serious in believing that nitrate in water is a hazard to health (unlikely) or that it is harming the environment (probable) we need to ask what we can do to help, individually and corporately.

We can, for example, waste less water. If the water companies need to build expensive denitrification plants or pipelines to meet the EC limit for nitrate by treatment or blending, it will make their job easier if we consume less of this increasingly scarce resource. We treat water in an even more cavalier fashion than energy. Only a fraction of the water we use actually passes into our bodies or touches our food. This is worth remembering next time you leave the tap running all day or wash the car or fill the bath luxuriously to the brim. Some have suggested that there should be two different water supplies, one for industry, washing or the lavatory, and another for drinking or cooking. It has even been suggested that the drinking water might be supplied from standpipes to get round the obvious problem of routing two pipelines to every house. Two water supplies would probably be unacceptable both economically to the water companies and politically to the government, but it raises the question of the bottled waters many people now buy from supermarkets. Most of these are relatively low in minerals despite the name 'mineral water'. According to a recent report in *Which* magazine only one brand of bottled water out of several tested contained amounts of nitrate anywhere near the EC limit for drinking water. They did, however, all contain very many more bacteria than tap water and so they must be boiled before being supplied to infants – the one human group proven to be at risk from nitrate.

If constructing two water *supply* lines is an unacceptable solution what about two sewage conduits? This may be more feasible. The problem, as the reader will recall from Chapter 8 is that nitrogen-rich domestic sewage mixes with industrial effluent containing heavy metals and other toxic substances that cannot be spread on land. If the domestic sewage could be intercepted before it mixed with industrial effluent, it could be composted to destroy pathogenic organisms and spread safely on farmland.

What else can we do? We could eat less meat. A lot of the grain grown in the world is fed to animals. Pigs and poultry in the Western World are almost exclusively reared in confined, ungrassed quarters and fed from arable produce. The pork and chicken meat then feeds another animal – humans. Animals incorporate protein from another source in a rather wasteful fashion; it would be about six or seven times more efficient if humans were to obtain protein directly from plant produce rather than through the intermediate of meat. Research has shown that it is possible to extract proteins directly from plants, but the product is not particularly appetizing and has proved more popular in the wet, tropical parts of the world, where there are plenty of suitable plant sources and plenty of hungry people, than in the well-fed Western World. Ruminants such as cattle or sheep remain by far the most palatable way we have of converting grass into a form which humans can digest. Pigs are well known for their habit of eating scraps and waste, but it makes more sense to farm these animals where the waste is produced rather than feed them with grain grown on land that could feed us directly. During and after the Second World War, pigs were raised very satisfactorily on the remains of school dinners and similar materials. It would be to everyone's advantage if we could return to this and other forms of recycling on a larger scale *without* the incentive of a war.

Action on nitrates

We can summarize most of what has been said in this chapter with a set of guidelines for farming practice. Put simply, anything that lessens the amount of nitrate at risk in the soil during autumn and winter helps to cut the leakage:

* nitrogen fertilizer should not be applied in autumn
* wherever possible the soil should not be left bare over winter (sow a catch-crop if necessary)
* sow winter crops early in autumn
* use animal manures only intermittently
* avoid ploughing-in old grassland
* plough-in straw
* use nitrogen fertilizer responsibly in accordance with professional advice. It may be safer to 'under-fertilize' rather than 'over-fertilize'

Each individual consumer of water and farm products will have to make up his or her own mind whether or not nitrate is a matter for concern to health. We think not. After all, it is estimated that water containing nitrate just above the EC limit contributes only about 50% of our daily intake (e.g. Dudley, 1986). But if you wish to err on the side of safety then we recommended the following:

* avoid large intakes of green leafy vegetables
* check processed meat products for nitrite and nitrate (E247–251)
* find out where your water comes from and ask the water company what the nitrate concentration is. If you wish to drink less nitrate, you could take part of your water intake from bottled mineral waters, but boil them before using them to make up in baby-feeds

EC law requires that the water companies supply water containing no more nitrate than 11.3 mg l^{-1}. They will have to comply. Nitrate in the general environment is a far larger risk because it makes some things grow too well at the expense of others. In the end it will be our wastefulness that aggravates this problem and probably not that of farmers.

The search for solutions to the nitrate problem that are both appropriate and effective has depended on identifying its real causes. We hope to have shown you that the nitrate problem is far more complex than is often thought and that it is not only farmers who are responsible for it. Politicians have contributed to the problem, so have pigeons, and so have you and I. We hope too to have shown why some simple and superficially attractive solutions probably will not work. If we wish to cut the amounts of nitrate in food, water and the environment we must pay attention to all the processes of the nitrogen cycle. Finally, we must take a holistic view of the environment and ensure that in countering the nitrate problem we do not simply exchange it for an equally complex alternative environmental problem.

Bibliography

Acheson, E.D. (1985) *Nitrate in Drinking Water.* HMSO, London, CMO (85), 14.

Addiscott, T.M. (1977) A simple computer model for leaching in structured soils. *Journal of Soil Science* **28**, 554–563.

Addiscott, T.M. (1983) Kinetics and temperature relationships of mineralization and nitrification in Rothamsted soils with differing histories. *Journal of Soil Science* **34**, 343–353.

Addiscott, T.M. (1988) Long-term leakage of nitrate from bare, unmanured soil. *Soil Use and Management* **4**, 91–95.

Addiscott, T.M. and Darby, R.J. (1991) Relating the nitrogen fertilizer needs of winter wheat crops to the soil's mineral nitrogen. Influence of the downward movement of nitrate in winter and spring. *Journal of Agricultural Science, Cambridge* **117**, 241–249.

Addiscott, T.M. and Powlson, D.S. (1989) Laying the ground rules of nitrate. *New Scientist* 29 April.

Addiscott, T.M. and Powlson, D.S. (1992) Partitioning losses of nitrogen fertilizer between leaching and denitrification. *Journal of Agricultural Science* **118**, 101–107.

Addiscott, T.M. and Wagenet, R.J. (1985) Concepts of solute leaching in soils: a review of modelling approaches. *Journal of Soil Science* **36**, 411–424.

Addiscott, T.M. and Whitmore, A.P. (1987) Computer simulation of changes in soil mineral nitrogen and crop nitrogen during autumn, winter and spring. *Journal of Agricultural Science* **109**, 141–157.

Addiscott, T.M. and Whitmore, A.P. (1991) Simulation of solute leaching in soils of differing permeabilities. *Soil Use and Management* **7**, 94–102.

Al-Dabbagh, S., Forman, D., Bryson, D., Stratton, I. and Doll, R. (1986) Mortality of nitrate fertilizer workers. *British Journal of Industrial Medicine* **43**, 507–515.

Alexander, J.D. (1987) Nitrogen-loss potential ratings for Illinois soils. US Department of Agriculture Bulletin **784**.

Aller, L., Bennet, T., Lehr, J.H. and Petty, R.J. (1985) *DRASTIC: A Standardized System for Evaluating Groundwater Pollution Potential using Hydrogeological Settings: USEPA* No. 600/2-85/018. United States Environmental Protection Agency, Ada Oklahoma, pp. 1–163.

Allison, F.E. (1955) The enigma of soil nitrogen balance sheets. *Advances in Agronomy* **7**, 213–250.

Barbee, G.C. and Brown, K.W. (1986) Comparison between suction and free-drainage soil solution samplers. *Soil Science* **141**, 149–154.

Belford, R.K. (1979) Collection and evaluation of large-scale monoliths for soil and crop studies. *Journal of Soil Science* 30, 363–373.

Beresford, S.A. (1985) Is nitrate in the drinking water associated with gastric cancer in the urban UK? *International Journal of Epidemiology* 14, 57–63.

Beven, K. (1981) Micro-, meso- and macro-porosity and channelling phenomena in soils. *Soil Science Society of America Journal* 45, 1245.

Bezdicek, D.F., Power, J.F., Keeney, D.R. and Wright, M.J. (1984) Organic farming: Current technology and its role in a sustainable agriculture. *American Society of Agronomy Special Publication* 46, ASA Madison, WI, pp. 1–36.

Boyce, J.S., Muir, J., Edwards, A.P., Seim, E.C. and Olson, R.A. (1976) Geologic nitrogen in Pleistocene loess of Nebraska. *Journal of Environmental Quality* 5, 93–96.

Briggs, L.J. and McCall, A.G. (1904) An artificial root for inducing capillary movement of soil moisture. *Science* 20, 566–569.

Brink, N. and Linden, B. (1980) Where does the commercial fertilizer go? *Eko Hydrologi* 7, 3–20.

Burns, I.G. (1974) A model for predicting the redistribution of salts applied to shallow soils after excess rainfall or evaporation. *Journal of Soil Science* 25, 165–178.

Burns, I.G. (1975) An equation to predict the leaching of surface applied nitrate. *Journal of Agricultural Science*, 85, 443–454.

Busch, D. and Meyer, M. (1982) A case of infantile methaemoglobinaemia in South Dakota. *Journal of Environmental Health* 44, 310.

Cannell, R.Q., Goss, M.J., Harris, G.L., Jarvis, M.G., Douglas, J.T., Howse, K.R. and Le Grice, S. (1984) A study of mole-drainage with simplified cultivation for autumn-sown crops on a clay soil. 1 Background, experiment and site details, drainage systems, measurement of drainflow and summary of results. *Journal of Agricultural Science* 102, 539–559.

Chaney, K. (1990) Effect of nitrogen fertilizer rate on soil nitrogen content after harvesting winter wheat. *Journal of Agricultural Science* 114, 171–176.

Choquette, K. (1980) Nitrates: Groundwater. *Journal of the Iowa Medical Society* 70, 309–311.

Christian, D., Goss, M.J., Howse, K.R., Powlson, D.S. and Pepper, T.J. (1990) Leaching of nitrate through soil. *IACR Report for 1989*, Lawes Agricultural Trust, 67–68.

Clough, P.W.L. (1983) Nitrates and gastric carcinogenesis. *Minerals and the Environment* 5, 947–959.

Cobbett, W. (1830) RURAL RIDES in the counties of Surrey, Kent, Sussex, Hampshire, Wiltshire, Gloucestershire, Herefordshire, Worcestershire, Somersetshire, Oxfordshire, Berkshire, Essex, Suffolk, Norfolk and Hertfordshire: WITH Economical and Political Observations relative to matters applicable to, and illustrated by, the state of those Counties respectively. London, published by William Cobbett, 183, Fleet Street.

Coleman, E.A. (1946) A laboratory study of lysimeter drainage under controlled soil moisture tension. *Soil Science* 62, 365–382.

Comly, H.H. (1945) Cyanosis in infants caused by nitrates in well-water. *Journal of the American Medical Association* 129, 112–116.

Council of the European Communities (1980) Relating to the quality of water intended for human consumption. Council directive of 15 July 1980 80/778/ EEC. *Official Journal of the European Communities* No. L229/11–29.

Croll, B.T. and Hayes, C.R. (1988) Nitrate and water supplies in the United Kingdom. *Environmental Pollution* 50, 163–187.

Currie, J.A. (1961) Gaseous diffusion in the aeration of aggregated soils. *Soil Science* 92, 40–45.

Dagan, G. and Bresler, E. (1979) Solute dispersion in unsaturated heterogeneous soil at field scale. *Soil Science Society of America Journal* 43, 461–472.

Darby, R.J., Hewitt, M.V., Penny, A., Johnston, A.E. and McEwen, J. (1988) The effects of increasing length of ley on the growth and yield of winter wheat. *Rothamsted Report for 1987* Part I, 101–102.

Deak, S. (1985) quoted in WHO (1985). Health hazards from nitrates in drinking water. Report on a WHO meeting. Copenhagen 5–9 March, 1984.

De Smedt, F., Wauters, F. and Sevilla, J. (1986) Study of tracer movement through unsaturated sand. *Geoderma* 38, 223–236.

Dilz, K., Darwinkel, A., Boon, R. and Verstraeten, L.M.J. (1982) Intensive wheat production as related to nitrogen fertilization, crop protection and soil nitrogen: experience in the Benelux. *Proceedings of the Fertilizer Society* 211, 93–124.

Dollard, G.J., Atkins, D.H.F., Davies, T.J. and Healy, C. (1987) Concentrations and dry deposition velocities of nitric acid. *Nature* 326, 481–483.

Dowdell, R.J., Webster, C.P., Hill, D. and Mercer, E.R. (1984) A lysimeter study of the fate of fertilizer nitrogen in spring barley crops grown on shallow soil overlying chalk: crop uptake and leaching losses. *Journal of Soil Science* 35, 169–181.

Dudley, N. (1986) *Nitrates in Food and Water.* The London Food Commission, p. 72.

England, C.B. (1974) Comments on 'A technique using porous cups for water sampling at any depth in the unsaturated zone' by Warren W Wood. *Water Resources Research* 10, 1049.

Ewing, M.C. and Mayon-White, R.M. (1951) Cyanosis in infancy from nitrates in drinking water. *Lancet* 260, 931–934.

Forman, D., Al-Dabbagh, A. and Doll, R. (1985) Nitrate, nitrite and gastric cancer in Great Britain. *Nature* 313, 620–625.

Foster, S.S.D., Bridge, L.R., Greake, A.K., Lawrence, A.R. and Parker, J.M. (1986) The groundwater nitrate problem. *Hydrogeological Report* 86/2. British Geological Survey, Wallingford.

France, J. and Thornley, J.H.M. (1984) *Mathematical Models in Agriculture.* Butterworths, London.

Goss, M.J., Colbourn, P., Harris, G.L. and Howse, K.R. (1988) In D.S. Jenkinson and K.A. Smith (eds), *Nitrogen Efficiency in Agricultural Soils,* pp. 269–282. Elsevier Applied Science, Barking.

Goulding, K.W.T. (1990) Nitrogen deposition to arable-land from the atmosphere. *Soil Use and Management* 6, 61–63.

Greenwood, D.J., Verstraeten, L.M.J., Draycott, A. and Sutherland, R.A. (1987) Response of winter wheat to N-fertiliser: Dynamic model. *Fertilizer Research* 12, 139–156.

Haines, B.L., Waide, J.B. and Todd, R.L. (1982) Soil solution nutrient concentrations sampled with tension and non-tension lysimeters: Report of discrepancies. *Soil Science Society of America Journal* **46**, 658–661.

Hansen, E.A. and Harris, A.R. (1975). Validity of soil–water samples collected with porous ceramic cups. *Soil Science Society of America Proceedings* **39**, 528–536.

Hansen, J.F. (1989) Nitrogen balance in agriculture in Denmark and ways of reducing the loss of nitrogen. In J.C. Germon (ed.), *Management Systems to Reduce Impact of Nitrates.* Elsevier, London.

Hansen, S. and Aslyng, H.C. (1984) Nitrogen balance in Crop Production Simulation model NITCROS. Royal Veterinary and Agricultural University, Copenhagen, 1984.

Harris, G.L., Goss, M.J., Dowdell, R.J., Howse, K.R. and Morgan, P. (1984) A study of mole-drainage with simplified cultivation for autumn-sown crops on clay soil. 2. Soil water regimes, water balances and nutrient loss on drain water. *Journal of Agricultural Science* **102**, 561–581.

HMSO (1975) *Food From Our Own Resources.* Cmnd 6020, London.

House of Lords Select Committee on the European Communities (1989) *Nitrate in Water.* HMSO, London.

Hye-Knudsen, P. (1985) Nitrate in drinking water and methaemoglobinaemia. *Ukeskr. Laegr* **141**, 51–53.

International Atomic Energy Authority (1984) *Soil and Fertilizer Nitrogen.* Technical Report **244** IAEA, Vienna.

Jaakkola, A. (1984) Leaching losses of nitrogen from a clay soil under grass and cereal crops in Finland. *Plant and Soil* **76**, 59–66.

Jarvis, S.C., Barraclough, D., Williams, J. and Rook, A.J. (1991) Patterns of denitrification loss from grazed grasslands: Effects of N fertilizer inputs at different sites. *Plant and Soil* **131**, 77–88.

Jarvis, S., Barraclough, D., Unwin, R.J. and Royle, S.M. (1989) Nitrate leaching from grazed grassland and after straw incorporation in arable soils. In J.C. Germon (ed.) *Management Systems to Reduce Impact of Nitrates.* Elsevier, London, pp. 110–125.

Jenkinson, D.S. (1985) How straw incorporation affects the nitrogen cycle. In J. Hardcastle (ed.) *Straw, Soils and Science.* AFRC, London, pp. 14–15.

Jenkinson, D.S. (1986) Nitrogen in UK arable agriculture. *Journal of the Royal Agricultural Society of England* **147**, 178–189.

Jenkinson, D.S. (1988) The turnover of organic matter in soils. In A. Wild (ed.), *Russell's Soil Conditions and Plant Growth*, 5th edn., 564–607. Longman, London.

Jenkinson, D.S. (1990) An introduction to the global nitrogen cycle. *Soil Use and Management* **6**, 56–61.

Jenkinson, D.S. and Johnston, A.E. (1977) *Soil Organic Matter in the Hoosfield Continuous Barley Experiment.* Rothamsted Report for 1976, part 2, pp. 87–101.

Joffe, J.S. (1932) Lysimeter studies. I. Moisture percolation through the soil profile. *Soil Science* **34**, 123–143.

Johnson, C.J., Bonrud, P.A., Dosch, T.A., Kilness, A.W., Serger, K.A., Busch, D.C. and Meyer, M.R. (1987) Fatal outcome of methaemoglobinaemia in an infant. *Journal of the American Medical Association* **257**, 2796–2797.

Jones, R.J.A., Thomasson, A.J., Robson, J.D. and Carter, A.D. (1989) The distribution of potential nitrate leaching losses from agricultural land in the UK based on computerized soil and climatic data. In R.J.A. Jones and B. Biagi (eds), *Agriculture, Computerization of Land Use Data.* CEC.

Jury, W.A. (1982) Simulation of solute transport using a transfer function model. *Water Resources Research* **18**, 363–368.

Keeney, D.R. (1989) Sources of nitrate to ground water. In R.F. Follet (ed.), Nitrogen management and ground water protection. *Developments in Agricultural and Managed-forest Ecology* **21**, Elsevier, Amsterdam, pp. 23–34.

Kolle, W., Strebel, O. and Bottcher, J. (1985) Formation of sulphate by microbial denitrification in a reducing aquifer. *Water Supply Berlin 'B'* **3**, 35–40.

Laurence Gould Consultants Ltd (1985) *Reduction of nitrate in water supplies: An economic study of water treatment compared with the effects of lower fertilizer usage.* DoE, Crown Copyright.

Lawes, J.B., Gilbert, J.H. and Warington, R. (1882) On the amount and composition of drainage water collected at Rothamsted. III. The quantity of nitrogen lost by drainage. *Journal of the Royal Agricultural Society of England,* 2nd Series **18**, 43–71.

MacDonald, A.J., Poulton, P.R. and Powlson, D.S. (1990) Arable crops and farming practices: Effects on nitrate in arable soils. In A. Scaife (ed.), *Proceedings of the First Congress of the European Society of Agronomy.* ESA, Session 5, p. 28.

Macdonald, A.J., Powlson, D.S., Poulton, P.R. and Jenkinson, D.S. (1989) Unused fertilizer nitrogen and its contribution to nitrate leaching. *Journal of the Science of Food and Agriculture* **46**, 407–419.

Marks, H.F. (1989) In D.K. Britton (ed.), *A Hundred Years of British Food and Farming. A Statistical Survey.* Taylor and Francis, London.

McGill, W.B., Hunt, H.W., Woodmansee, R.G., Reuss, J.O. and Paustian, K.H. (1981) Formulation, process controls, parameters and performance of PHOENIX: a model of carbon and nitrogen dynamics in grassland soils. In M.J. Frissel and J.A. van Veen (eds), *Simulation of Nitrogen Behaviour of Soil–Plant Systems.* Pudoc, Wageningen, pp. 171–191.

Medawar, P.B. (1967) *The Art of the Soluble.* (See especially the essay 'Hypothesis and Imagination'.) Methuen, London.

MAFF (1990) *Agriculture in the United Kingdom: 1989.* HMSO, London.

MAFF (1947–1990) *Agricultural Statistics United Kingdom.* HMSO, London.

MAFF (1971) *The Significance of Winter Rainfall over Farmland in England and Wales,* Technical Bulletin **24**. HMSO, London.

MAFF (1951) Agriculture: The Ploughed-up Grassland (Fertilizers) Scheme, 1951. *Statutory Instruments* 1153.

Molina, J.A.E., Clapp, C.E., Shaffer, M.J., Chichester, F.W. and Larson, W.E. (1983) NCSOIL, a model of nitrogen and carbon transformations in soil: description, calibration and behaviour. *Soil Science Society of America Journal* **47**, 85–91.

Muller, J.C., Denys, D., Morlet, G. and Mariotti, A. (1989) Influence of catch crops on mineral nitrogen leaching and its subsequent plant use. In J.C. Germon (ed.), *Management Systems to Reduce Impact of Nitrates.* Elsevier, London, pp. 85–98.

Parkin, T.B. (1987) Soil microsites as a source of denitrification variability. *Soil*

Science Society of America Journal **51**, 1194–1199.

Philip, J.R. (1991) Soils, natural science and models. *Soil Science* **151**, 91–98.

Popper, K.R. (1959) *The Logic of Scientific Discovery.* London, Hutchinson.

Power, J.F. and Schepers, J.S. (1989) Nitrate contamination of groundwater in North America. *Agriculture, Ecosystems and Environment* **26**, 165–187.

Powlson, D.S., Pruden, G., Johnston, A.E. and Jenkinson, D.S. (1986) The nitrogen cycle in the Broadbalk wheat experiment: recovery and losses of ^{15}N-labelled fertilizer applied in spring and impact of nitrogen from the atmosphere. *Journal of Agricultural Science, Cambridge* **107**, 591–609.

Powlson, D.S., Brookes, P.C. and Christensen, B.T. (1987) Measurement of soil microbial biomass provides an early indication of changes in total soil organic matter due to straw incorporation. *Soil Biology and Biochemistry* **19**, 154–164.

Powlson, D.S., Poulton, P.R., Addiscott, T.M. and McCann, D.S. (1989) Leaching of nitrate from soils receiving organic or inorganic fertilizers for 135 years. In J.A. Hansen and K. Hendricksen *Nitrogen in Organic Wastes Applied to Soils*, pp. 334–345. Academic Press, London.

Powlson, D.S., Hart, P.B.S., Poulton, P.R., Johnston, A.E. and Jenkinson, D.S. (1992) The influence of soil type, crop management and weather on the recovery of ^{15}N-labelled fertilizer applied to winter wheat in spring. *Journal of Agricultural Science, Cambridge* **118**, 83–100.

Prins, W.H., Dilz, K. and Neeteson, J.J. (1988) Current recommendations for nitrogen fertilization within the EEC in relation to nitrate leaching. *Proceedings of the Fertilizer Society* **276**, 27pp.

Ramos, C., Domingo, R. and Oliver, J. (1989) Nitrate leaching under two nitrogen fertilization managements. In J.C. Germon (ed.), *Management Systems to Reduce Impact of Nitrates.* Elsevier, London, pp. 99–109.

Recous, S., Fresneau, C., Faurie, G. and Mary, B. (1988) The fate of ^{15}N-labelled urea and ammonium nitrate applied to a winter wheat crop: II Plant uptake and N efficiency. *Plant and Soil* **112**, 215–224.

Richards, L.A., Neal, O.R. and Russell, M.B. (1939) Observations on moisture conditions in lysimeters, II. *Soil Science Society of America Proceedings* **4**, 55–59.

Richter, J., Kersebaum, K.C. and Nordmeyer, H. (1985) Simulation of nitrogen regime in loess soils in the winter half year: comparison between field measurements and simulations. *Plant and Soil* **83**, 419–431.

Risch, H.A., Jain, M. and Choi, N.W. (1985) Dietary factors and the incidence of cancer of the stomach. *American Journal of Epidemiology* **122**, 947–959.

The Royal Society (1983) *The Nitrogen Cycle of the United Kingdom: A Study Group Report.* The Royal Society, London.

Russell, A.E. and Ewel, J.J. (1985) Leaching from a tropical Andept during big storms: a comparison of three methods. *Soil Science* **139**, 181–189.

Russell, E.W. (1973) *Russell's Soil Conditions and Plant Growth*, 4th edn, p. 2. Longman, London.

Sattelmacher, P.G. (1962) Methaemoglobinaemia durch nitrate in Trinkwasser. *Schriflenreie Verein*, Wasser Boden Lufthyg. Berlin-Dahlen **20**, Gustav Fischer Verlag, Stuttgart.

Shaffer, K.A., Fritton, D.D. and Baker, D.E. (1979) Drainage water sampling in a wet dual-pore system. *Journal of Environmental Quality* **8**, 241–246.

Shepherd, M. (1990) Measuring soil N is worthwhile in some situations. *Arable Farming* November, pp. 39 and 43.

Sherwood, M. and Fanning, A. (1989) Leaching of nitrate from simulated urine patches. In J.C. Germon (ed.), *Management Systems to Reduce Impact of Nitrates.* Elsevier, London, pp. 32–44.

Smith, J.H. and Peterson, J.R. (1982) Recycling of nitrogen through land application of agricultural, food processing and municipal wastes. In F.J. Stevenson (ed.), *Nitrogen in Agricultural Soils.* American Society of Agronomy monograph 22, ASA, Madison, Wisconsin.

Smith, K.A. (1980) A model of the extent of anaerobic zones in aggregated soils, and its potential application to estimates of denitrification. *Journal of Soil Science* 31, 263–277.

Smith, K.A. and Unwin, R.J. (1983) Fertilizer value of organic manures in the UK. *Proceedings of the Fertilizer Society* 221, 31pp.

Smith, K.A., Elmes, A.E., Howard, R.S. and Franklin, M.F. (1984) The uptake of soil and fertilizer nitrogen by barley growing under Scottish climatic conditions. *Plant and Soil* 76, 49–57.

Stanford, G. and Epstein, E. (1974) Nitrogen mineralization–water relationships in soils. *Soil Science Society of American Proceedings* 38, 103–107.

Stanford, G. and Smith, S.J. (1972) Nitrogen mineralization potentials of soils. *Soil Science Society of America Proceedings* 36, 465–472.

Strathouse, S.C., Sposito, G., Sullivan, P.J. and Lund, L.J. (1980) Geologic nitrogen: A potential geochemical hazard in the San Joaquin Valley, California. *Journal of Environmental Quality* 9, 54–60.

Sylvester-Bradley, R., Addiscott, T.M., Vaidyanathan, L.V., Murray, A.W.A. and Whitmore, A.P. (1987) Nitrogen advice for cereals: Present realities and future possibilities. *Proceedings of the Fertilizer Society* 263, 36pp.

Tabatabai, M.A. and Al-Khafaji, A.A. (1980) Comparison of nitrogen and sulphur mineralization in soils. *Soil Science Society of America Journal* 44, 1000–1006.

Thornley, J.H.M. and Verberne, E.L.J. (1989) A model of nitrogen flows in grassland. *Plant Cell and Environment* 12, 863–886.

UK DoE (1986) *Nitrate in Water.* Department of the Environment Pollution Paper 26, HMSO, London.

Van Burg, P.F.J., Prins, W.H. den Boer, D.J. and Sluiman, W.J. (1981) Nitrogen and intensification of livestock farming in EEC countries. *Proceedings of the Fertilizer Society* 199, 78pp.

Van der Ploeg, R.R. and Beese, F. (1977) Model calculations for the extraction of soil water by ceramic cups and plates. *Soil Science Society of America Journal* 41, 466–470.

Wagenet, R.J. (1983) Principles of salt movement in soils. In D.W. Nielsen *et al.* (eds), *Chemical Mobility and Reactivity in Soil Systems,* Special publication 11. American Society of Agronomy, Madison, WI.

Wagenet, R.J. and Rao, B.K. (1983) Description of nitrogen movement in the presence of spatially variable soil hydraulic properties. *Agricultural Water Management* 6, 227–242.

Wagner, G.H. (1962) Use of porous ceramic cups to sample soil water within the profile. *Soil Science* 94, 379–386.

Wehrmann, J. and Scharpf, H-C. (1989) Reduction of nitrate leaching in a vegetable farm: fertilization, crop rotation, plant residues. In J.C. Germon (ed.), *Management Systems to Reduce Impact of Nitrates.* Elsevier, London, pp. 147–157.

Which (1989) February, p. 58.

Whitmore, A.P. and Addiscott, T.M. (1987) Applications of computer modelling to predict mineral nitrogen in soil and nitrogen in crops. *Soil Use and Management* **3**, 38–43.

Whitmore, A.P., Bradbury, N.J. and Johnson, P.A. (1992) The potential contribution of ploughed grassland to nitrate leaching. *Agriculture, Ecosystems and Environment* **39**, 221–233.

Widdowson, F.V., Penny, A., Darby, R.J., Bird, E. and Hewitt, M.V. (1987) Amounts of NO_3-N and NH_4-N in soil, from autumn to spring, under winter wheat and their relationship to soil type, sowing date, previous crop and N uptake at Rothamsted, Woburn and Saxmundham, 1979–85. *Journal of Agricultural Science, Cambridge* **108**, 73–95.

Wong, M.F.T., Wild, A. and Juo, A.S.R. (1987) Retarded leaching of nitrate measured in monolith lysimeters in south-east Nigeria. *Journal of Soil Science* **38**, 511–518.

Wood, W.W. (1973) A technique using porous cups for water sampling at any depth in the unsaturated zone. *Water Resources Research* **9**, 486–488.

Index

Abuse of models 74
Abstraction of water 134, 140
Acacia 53
Acetylene 53
Acidity 32, 51
Active growth (of crop) 25, 94
Aerosols 128, 129
Agar 45
Aggregation (of soil) 39, 41, 42
Agricultural Development and Advisory
  Service (ADAS) 106
Agriculture
  abandonment of 138, 139
  controls on 138
  extensification 29, 117
  intensification 29, 117
Air–water interface 40, 67, 70
Alder 53
Alfalfa (lucerne) 53
Algae 7
  blue-green 52
  marine 14, 136
  toxic effects 7
Alkalinity 32, 34, 51
Aluminium 38
Amine, secondary 9
Ammonia 32, 34, 117, 128, 132
Ammonification 46
Ammonium 15, 24, 35, 46, 47, 146
  ion 32, 33
Ammonium nitrate 24, 25, 35
Animals (farm) 117
  feed 118
    contribution to nitrate problem 118
    organic 123
    use of straw in 149
  manures see Manures
  methane production 149
  slurry see Slurry
Aquatic plants 134
Aquifers 3, 135, 136
  chalk 1
  confined and unconfined 134
  denitrification in 134
  downward movement in 136

fissuring in 136
limestone 136
nitrate in 135, 136
residence time 135, 136
sandstone 1
Arrhenius equation 87
Artificial root 56
Atomic mass 97
Autumn 21, 93, 94, 102, 103
  fertilizer applications 26, 94, 155
Baby feeds 155
Bacteria
  in mouth 11
  in soil 45
  in stomach 7
  nitrifying 24, 46, 47
Bacterial colonies 45
Balance of payments 5, 139
Bare soil 30, 59, 92, 94, 102, 139, 144, 155
Barley 17
Beans (field) 16, 53, 123
Bentonite 57
Best buy (of methods for measuring
  leaching) 71–2
Biomass 44, 48
Bicarbonate 133
Blood
  dried 47, 48
  oxygenation of 7
Blue-baby syndrome 7, 8
Brassicas 144
Brimstone experiment 69–7, 102
British government 13, 109, 121, 140, 153
Broadbalk Field 105, 106, 107
  nitrate in drainage 106, 107
Boreholes 3, 128, 136, 140
Botulism 6, 12
Bulk density (of soil) 37
By-pass flow (of water) 43

Cadmium 59, 61
Calcium 15, 17, 31, 33
Calcium nitrate 25, 33, 35
Cambridgeshire 113
Cancer 9, 10, 11

Capacity
　of soil to hold water　43
　parameters in models　75–80
Capillarity　40, 43
Carbohydrate　16, 17
Carbon　16, 52
Carbon dioxide　15, 16, 46, 52
Carbon disulphide　34
Catch crops (cover crops)　103, 125, 126,
　140, 144, 145, 155
　phacelia　145
　radish　144, 145
　rape　145
　rye　144, 145
　turnips　144
　use of nitrogen　145, 146
　use of water　144
　volunteers　145
　white mustard　144, 145
Cattle　117, 119
Cells　15, 18
Ceramic cups *see* Porous ceramic cups
Channelling (of water flow)　43, 57
Charcoal　33
Chemistry　31–7
Chemical potential　37
Chilean nitrate　123
Classical experiments　128, 137
Clay　37, 39, 41, 46, 47
　cracking and swelling　38
Climate change　51, 130
Clover　16, 53, 54, 118
C:N ratio　48
Collector (for water)　59, 62, 65, 67
Common Agricultural Policy　4
Combustion　129
Convection-dispersion equation　81–3
Consumer, role in nitrate problem　109, 153,
　154, 155
Corrosion　133
Cracks　38, 43
Crop development　18
Crop protection　19, 21
Crop quality　18, 125
Cultivation　39, 92
Cured meat　12 *see also* Preserved meat
Cyanide　53

DCD　34
Demand, for nitrogen by crop　25
Denitrification　48, 49, 96, 99, 100, 130, 132,
　133, 134
　plants　14, 133, 154
　potential　50
Diffusion　36, 42, 43, 82
Dig for Victory campaign　3, 109
Disease(s)　19
Dispersion　42, 82
Dissociation　31

DNA　16
Drains　38, 69
　flow in　69, 70, 71
　mole　69
　mole and pipe　69, 70
　tile　69
Drain Gauges　62, 63, 80, 82, 104, 105
　nitrate losses　104, 105
Drinking water　1, 154
Dust　129
Dung　117, 119, 148

E247, E251　155
Early sowing (of winter crops)　93, 142, 143,
　155
Earthworms　46
East Anglia　134
Ebermeyer lysimeter　64, 65, 68
Education　140–1
Electric fences　147
Electrical charge　31, 34, 47
Electron acceptors　50
Energy　15, 39, 53, 96
Environmental Protection Agency　140
European community　1, 4, 130, 133, 153
　legislation　137
　nitrate limit　1, 13, 105, 116, 121, 128, 155
Eutrophication　7
Evaporation　102, 103

Famine　1, 6, 109
Farmers　3, 4
　blame for nitrate problem　5, 7, 153
　income support　138
　profit　108, 140
Farming
　extensive and intensive　29, 117
　good practice　140, 155
Fertilizer
　nitrogen
　　application　25, 44, 110, 116, 117, 155
　　direct losses from　43, 94, 104
　　financial return on　17, 21, 117
　　labelled *see* Labelled fertilizer
　　response to　17, 19, 20
Fibreglass pipe　65
Finite difference method　83
Finite element method　83
Fires　130
Fixation　52, 123, 150
Flow　40, 74, 82
　of water in soil　40–2, 76–82
　steady- and non-steady-state　82
Food
　prices　5, 17
　over-production　5
　shortages　1, 6, 109
Forestry　29, 92, 150, 153
　water use　134

Freezing of soil 34, 39, 140
Fuhrburg aquifer 133
Fungi 44, 46

Glaciers 132
Global warming 51, 130
Glucose 47, 48
Gradient (of potential etc.) 36
Grain
  fed to animals 154
  mountains 3, 5, 139
  price 21
  reserves 139
  yield 18
Grass–clover leys
  as a means of lessening nitrate
    leaching 147, 148
  contribution to nitrate leaching 118
  give free nitrogen 147
  in organic farming 123
Grassland 28, 92, 110–22, 147
  area under 110
  cut 116
  grazed 117, 118
  inefficient use of nitrogen 117
  nitrogen balance 117, 118
  permanent 112
  response to nitrogen 116, 117
  stocking rate 117, 125
    and productivity 117, 125
  surplus nitrate curve 111
  temporary 112, 113
  water use 134
Grazing 148
Greenhouse
  effect 51, 130
  gases 51, 149
Growth regulators 19

Haemoglobin 7
Hanging column (of water) 67, 72
Headlands 146
Hedge greens 146
Herring-bone cracks (in soil) 69
Hillslope 110
Hoosfield 129, 137
Horticulture 150
Hot spots (for denitrification) 50
House of Lords 1, 9
Humus 44
Hydraulic conductivity 43, 83
Hydraulic potential 83
Hydroxide 32
Hyphae (fungal) 46
Hypothesis 73, 74

Immobilization 47
Inhibitors (of nitrification) 34, 146

Interflow 69, 71
Ions 31
Ion exchange 14, 133
Iron 133
Irrigation 36, 42, 132, 138, 151

Labelled fertilizer 17, 97–102, 142
  fate 97
    oilseed rape 101
    potatoes 101
    winter wheat 98, 101
  loss of
    partitioned between leaching and
      denitrification 99, 100
    related to rain after application 99
Labour, supply of 127
Lakes 1
Land prices 3, 5
Lateral flow 65, 67, 70, 72
Leaching
  equation 76
  models *see* Models, leaching
  of nitrate *see* Nitrate, leaching
Leaf area 18
Legumes 16, 53, 123
Leucaena 53
Leys
  ploughing of 112
Lincolnshire 113
Lodging of crops 19, 123
Lucerne (alfalfa) 53
Lysimeter(s) 62–9
  Ebermeyer 64, 65, 67, 72
  monolith 64, 65, 66, 67, 72, 102

Magnesium 15, 33
Manure(s) 119
  application 120, 121, 155
  bulk 120
  composition 119, 120
  disposal 120, 140
  losses of nitrogen to atmosphere 121
  organic 126
  quantity produced (UK) 119
  runoff problem 136, 140
  storage 120, 139
  transport 120
  value
    manurial 119
    monetary 119
Maps 141
Mass flow 81
Mass spectrometer 97
Meat
  consumption 154
  production 154
    inefficient use of nitrogen 155
Mechanization 3

Memory effects 105-7
Meniscus 40
Methaemoglobinaemia
  infants 7, 8
  ruminants 117
Methanol 133
Microbes 7, 32, 35, 45, 46, 49, 74, 85, 103,
  133, 151
Microbial biomass 44, 48
Microcosms 47
Micro-organisms
  soil 45-50
Millipedes 46
Mineral water 154
  bacteria in 154
Mineralization 25, 46, 94, 110, 119, 144
  kinetics 86-9
    first-order 86
    zero order 86
Model, a definition 73
Models 73-91, 139
  capacity and rate 75
  complexity 75, 83, 85
  crop growth 89, 90
  deterministic *vs* stochastic 75
  estimation of nitrate leaching from soil
    sampling 71
  functional *vs* mechanistic 75
  leaching 75-84
    classical physical 81-3
    DRASTIC 140, 141
    functional 77-81
    mechanistic 81
    stochastic 84
  mineralization 85-9
  nitrogen uptake 89
  purpose 75
  validation 74
Modelle.s 74
Mortality 12
Motor vehicles 130

National Dried Egg 3
National Resource Inventory 141
Nitrate 24, 35
  assimilation (by plants) 15, 24
  chemical properties 31
  destruction by microbes 48, 49, 50
  geological 131, 132
  health hazards
    cancer of the stomach 9, 10, 11, 12
    methaemoglobinaemia 7
  in natural waters 1, 3
  in soil at harvest 97, 101
  in vegetables 10, 131
    effects of storage and freezing 131, 132
  in water supplies 134-6
  leaching 74, 94, 99, 100
    measurement 55-72

  reduction of 138, 141, 142, 155
  medical use 7
  production by microbes 33, 46, 103, 104
  relation to nitrite 6
  Sensitive Areas 5, 140
  solubility 32, 33
  uptake by crops 92, 93, 95
Nitrate reductase 24
Nitric acid 130
Nitrification 32, 46, 47, 51
Nitrite 24, 31, 46, 131
  health hazards
    cancer of the stomach 9, 10, 11, 12
    methaemoglobinaemia 7
  in preserved meat 12, 131
  in vegetables 131
    effects of storage and freezing 131, 132
  relation to nitrate 6, 131
Nitrite reductase 24
Nitrogen 16
  balance 48, 129, 141
  deficiency 17
  fertilizer 3, 50, 52, 95, 112, 116, 117
    autumn application 26, 94, 155
    placement 153
    price 17
    professional advice 155
    split application 153
  fixation 52, 123, 150
    biological 16, 52
    chemical 52
  gas 51, 130, 133
  mineral
    deposition from atmosphere 96, 128,
      130
    'flush' 86, 87
    in soil at harvest 97, 101
    pathways in soil 95
  organic 44, 76
  oxides 129, 130
  pools 85
Nitrogenase 53
Nitrobacter 47
Nitrosomonas 47
Nitrous oxide 14, 51, 96, 130, 133
N_{min} method 22, 151
Nodules, bacterial 16, 53
Northamptonshire 115
N-serve 34
Nucleic acids 16

Oilseed rape 19, 28, 144
  leaf shedding 105
  memory effect 105, 106
  residues at harvest 105, 108, 151
Oil price 139
Optimum (application of nitrogen fertilizer)
  economic 17, 19, 21, 22
  environmental 24

Organic farming 95, 123–7
  disturbance of ecosystems 123
  nitrate leaching 125, 126
    arable land 124, 125
    grassland 125
  nitrogen inputs 124
  productivity
    arable land 124
    grassland 125
  stocking rates 125
Organic food 123, 124
  premium 124
Organic matter 38, 39, 44, 95, 104, 105, 109, 153
Organic manures *see* Manures, organic
Over-production 22, 138, 139
Oxygen 7, 38, 49, 50, 53
Ozone 51, 130, 138

Parameter
  definition 75
Partial pressure 36
Peas 16, 53
Permanent grass
  ploughing of 109, 112–15
    contribution to nitrate problem 109, 114
Pesticides 123
pH 117
Phosphate 15, 31, 136
Phosphorus 16
Photosynthesis 15, 95
Physics 37–43
Pigeons 19, 28, 156
  droppings 139
  oilseed rape 19
Pigs 154, 155
Piston flow (piston displacement) 42, 76
Ploughing 112–15, 121, 125, 155
Probability 84
Poaching (of grassland by animals) 117, 147
Polders 38
Politicians 3, 109, 156
Pollution
  atmospheric 121
  farm accidents 148
  nitrate 17, 36
  silage effluent 148
Pores, soil 40, 41, 43, 58, 59, 67, 68
  suction 40, 43, 67, 68
Porous candles 68
Porous ceramic cups 56–62, 68, 69, 71
Porous ceramic plate 65, 68
Potassium 17, 31, 37
Potatoes 19, 92, 108, 152, 153
Potential energy 37
Pottery 38
Power stations 130

Predators 47, 147
Preserved meat 12, 131
Pressure 36, 40
Prey 47
Production 109
Protein(s) 46
Pyrites 133

Q_{10} 86
Quotas 147

Rain 36, 42, 59, 79, 102, 103, 112, 128
Rape *see* Oilseed rape
Rate parameters in models 75, 82–3
Ready reckoner 151
Recuperation (of land) 15
Reeds 6, 136
Reservoirs 134, 136
Residues 46, 108, 151, 153
Respiration 50
Response curve 19, 20
Rhizobia 16
Rice 28
Risk 25, 43
Rivers 1, 136
Rock phosphate 123
Root systems 153
Rooting depth 55, 150, 151
Rooting zone 55
Roots 16, 92, 95
  early establishment 93
  grass 92
Rotation of crops 15, 123, 139
Rough pasture 110

Saliva 9, 10
Sand 37, 38, 39
Seaweed 139
Seed-bed 39, 93
Set-aside 139, 140
Sewage
  discharge to sea 121, 135
  nitrogen content 122
  per person 134, 135
  processing 154
  sludge 122, 135
Sheep 29, 110
Silage 148
  pollution from 148
Silicon 38
Silt 37, 38
SLIM leaching model 79–81
Slurry 119
  application 120, 121
  bulk 120
  composition 119, 120
  losses of nitrogen to atmosphere 121
  quantity produced (UK) 119

runoff problem 136
transport 120
value
  manurial 119
  monetary 119
Snow 140
Spain 152, 153
Spinach 131, 142
Split applications (of fertilizer) 25, 26, 152
Spring 94
Springs 134
Springtails 46
Soil
  acidification 130
  aggregates 39, 42
    formation 38, 39
    size 39
    role 41
    stabilization 39
  effect of disturbance on water flow 63
  particles 37, 38, 56
    classification 37, 38
    clay 37, 39, 41, 46, 47
    silt 37, 39
    sand 37, 39
  ripening 38
  sampling 23, 71, 151
  structure 42, 58, 137
    and organic matter 39
    role in nitrate leaching 41, 42, 43
  texture 43
  water
    movement 40, 41
    saturation 40, 67
    unsaturation 40
Soils
  heavy 69, 70
  light 74, 121
Soil Survey and Land Resource Centre 43
Stomata 129
Straw
  carbon to nitrogen ratio in 48
  immobilization of nitrate by 141, 143
  ploughing in 155
  use in animal feed 149
Streams 134
Stubble burning 130
Subsidies 3, 140, 153
Subsoil 38, 39, 41
Sugar beet 19, 20, 92, 150
Sulphate 15, 17, 31, 132, 133
  problems 133
Surface tension 40, 67
Surface runoff 69, 71
Surplus nitrate curve 23, 105, 107, 111, 141, 153
  plateau 105, 107

Tap-water 1, 9, 138

Termites 46
Time lag (for cancer development) 11
Topping up (with fertilizer) 22
Travel time 84
Triple bond (of dinitrogen) 53
Trithiocarbamate 34
Tropics 28, 54, 59
Tunnel (for collecting soil water) 59, 65

UKROFS (United Kingdom Register of
  Organic Food Standards) 123
Urea 25, 35
Urease 25
Urine 117, 148, 149, 150

Variability 37, 43, 50, 57, 64, 67, 83, 124
Vegetables 10, 155
  brussels sprouts 150
  cabbage 150, 151
  cauliflower 150, 151
  celery 150
  french beans 150
  onions 152
  peas 150, 151
  spinach 131, 150, 151
Vitamin C 10, 12, 131
Vitamins 124
Volatilization of ammonia 128
Volumetric moisture content 43, 76, 77, 79, 82, 83

Water
  abstraction 134, 140
  drinking and cooking 1, 154
  flow in soil 40, 41, 72, 82
  in plants 15
  level 134
  mobile and immobile in soil 41, 42, 43, 55, 58, 63, 79, 81
  standpipes 154
  wastage 154
Water companies 134
Weather 9
Weeds 126
Wells 8
Wheat 17
  winter
    memory effects 105–7
    mineral nitrogen left at harvest 97
    percent of arable acreage 108
    residues 108
Winter 93
Work-force, agricultural 3, 4, 127, 139
World Health Organization (WHO) 13, 137
World population 123, 139
World War II 25

Yield(s) 18, 137